Space, the Final Frontier?

What are our motivations for going into space? Where does our long-term space future lie? Why, and how, should we strive to reach, if not for the stars, at least for the Moon and Mars? What are the greatest challenges, and advantages, of space to the human species?

This exciting book looks first at the progress that has already been made in our attempts to explore and expand beyond the Earth. Current and past space technologies and space stations are described, and the effects of the space environment on the human body are explained. A discussion of the merits of the robotic exploration of space is followed by a look at our exploration of the Moon and Mars, and exploitation of the solar system resources. Final chapters touch on propulsion methods required for leaving our solar system, and ask which of the possibilities for future space travel is most likely to succeed.

Whatever happens in the future, our expansion into space will change us deeply. This thought-provoking book will appeal to all those with an interest in the future of space exploration.

Giancarlo Genta is a Professor in the Department of Mechanics at the Technical University of Turin, Italy. He obtained degrees in Aeronautical Engineering and Aerospace Engineering from the same university, and his current research in applied mechanics is linked to the construction of machines. He is a corresponding member of the International Academy of Astronautics and of the Academy of Sciences of Turin. He has published eight previous books and a large number of research papers.

Michael Rycroft is a Visiting Professor at the International Space University in Strasbourg, France, and at De Montfort University in Leicester, UK. For 11 years he was Head of the Atmospheric Sciences Division at the British Antarctic Survey, Cambridge, and he also spent 5 years as Professor of Aerospace at Cranfield University. He has acted as Editor-in-Chief for the *Journal of Atmospheric and Solar-Terrestrial Physics*, is now Managing Editor of *Surveys in Geophysics*, and also edited the *Cambridge Encyclopedia of Space*, published by Cambridge University Press.

Space, the Final Frontier?

by
Giancarlo Genta
The Technical University of Turin, Italy

and
Michael Rycroft
The International Space University, Strasbourg, France
and De Montfort University, Leicester, UK

CAMBRIDGE
UNIVERSITY PRESS

PUBLISHED BY THE PRESS SYNDICATE OF THE UNIVERSITY OF CAMBRIDGE
The Pitt Building, Trumpington Street, Cambridge, United Kingdom

CAMBRIDGE UNIVERSITY PRESS
The Edinburgh Building, Cambridge CB2 2RU, UK
40 West 20th Street, New York, NY 10011-4211, USA
477 Williamstown Road, Port Melbourne, VIC 3207, Australia
Ruiz de Alarcón 13, 28014 Madrid, Spain
Dock House, The Waterfront, Cape Town 8001, South Africa

http://www.cambridge.org

First published 2003

Printed in the United Kingdom at the University Press, Cambridge

Typeface Trump Medieval 9.5/15 pt   *System* LaTeX 2_ε [TB]

A catalogue record for this book is available from the British Library

Library of Congress Cataloguing in Publication data

Genta, G. (Giancarlo)
Space, the final frontier / by Giancarlo Genta and Michael Rycroft.
    p. cm.
Includes bibliographical references and index.
ISBN 0 521 81403 0
1. Astronautics – Popular works.   2. Outer Space – Exploration – Popular
works.   I. Rycroft, Michael J.   II. Title.
TL793 .G426   2002
629.4–dc21   2002067234

ISBN 0 521 81403 0 hardback

To future generations of the human species

Contents

Foreword

by Franco Malerba

As the second millennium drew to a close, the event which most excited the imagination of humankind was landing men on the Moon. The *Apollo* astronauts had reached a distant, alien world, after leaving the planet to which they seemed to be constrained for ever, condemned to be born, to live and to die deep in its gravity well. They left their footprints on the Moon, still a mysterious celestial body, always considered to be beyond our reach, and described by poets for its magic. After that time of wonderful enterprises in space, so well described by the media, no other event was able to catch the public attention and arouse the interest of many people. But a time of rapid expansion of scientific knowledge about the Universe and of consolidation of space technology, followed. In the words of Dan Goldin, the previous NASA Administrator, '... we travelled to the beginning of time with the Hubble Space Telescope, discovered new galaxies and planets in other star systems, we captured the interest of the world with the discoveries of Pathfinder on the surface of Mars, we deepened our knowledge of planet Earth and of effects caused by the presence of humans...'. These are important results, even though today the media only occasionally report news on space research – that usually occurs only when some accident upsets the routine.

After twenty years of the Space Shuttle, and after *Mir*, the dawn of the twenty-first century marks the era of systematic exploration of the solar system and of the exploitation of near-Earth space with the *International Space Station*. A new season of new knowledge and technological progress is starting. Perhaps it will not be too long before humans from the blue planet land on the red planet.

'The exploration and exploitation of space and of celestial bodies are freely allowed to all countries, but neither planets nor space may be occupied or claimed... the exploration of space must be inspired by the principles of peaceful cooperation of peoples... all astronauts are representatives of humankind in space and have the right to receive all possible assistance in case of accidents or emergency...' So solemnly states the declaration of the Principles which govern the exploration of space as formulated by the United Nations Organisation and approved by its General Assembly. At the beginning of the third millennium, space became the new territory to be explored and conquered by humankind, a conquest which has an economic dimension, but also a moral and a cultural one.

To explore space means to know ourselves better. Consider that most enigmatic mystery of all, life. We can already see that the discovery of alien forms of life on some planet in the Universe will be a turning point. By comparing our knowledge of biology on the Earth with that on other worlds, we may better understand our own nature, in the same way as our own language is better understood when we learn a foreign one.

This book by Giancarlo Genta and Michael Rycroft deals with these themes. It leads us to the mysterious frontiers of the Universe, looking into possible space technologies, without giving way to fantasy, without setting limits to the imagination. What the authors provide is neither a historical account nor a scientific essay, but a conversation – understood even by the non-specialist reader – illustrating in turn various aspects of space exploration, space programmes already under way, and those which may follow.

This cradle is too small – to use a metaphor of the space pioneer Konstantin Tsiolkovsky – for humankind who has learned how to read the laws of the Universe and how to build machines: change is already under way. The need to learn and to explore, beyond all Pillars of Hercules, carries us on to new and more distant frontiers, to take our genes – and our values – into new realms.

Franco Malerba
First Italian astronaut

Foreword

by Michael Foale

Space, the Final Frontier – these are words that evoke the spirit and dreams that have spurred on countless children to think and wonder how to get themselves into the heavens. For me, an awareness of what space and exploration could be came at an early age, having been born in the same year that the first satellite, *Sputnik*, was launched by the former Soviet Union, in 1957. As the excitement of the race to put a person on the Moon was proclaimed by the world's media, it seemed only natural to me that young children such as myself should be able to grow up to explore Space, the Next Frontier.

Human history is extraordinary. Our technological and social progress over the last 10,000 years seems breathtakingly rapid, when compared with any development on Earth that preceded it. Geologic time, a measure of the Earth's history, is immense in comparison, similar to the lifetime of our galaxy, the Milky Way. If Earth's whole history were to take place in one day, humanity would appear only in the last 30 seconds of such a day, and our first steps into space would be no more significant than the blink of an eye. At the same time, we should not forget that our view is anthropocentric. It would be arrogant to deny the possibility that intelligent beings existed before us, and may again exist after us, in the course of the history of the Universe. From this perspective we might forgive ourselves for not having made greater strides away from the Earth! It is appropriate to the physical scale of the adventure to consider both the past and the future development of humans into space over a rather long time period.

A quality of people is to dream. Dreams give us a tremendous power to conceive of what might be possible, and have provided powerful motivations for people to strive to realise some of the great achievements made by human beings. Humanity made its debut in the last century, taking ageless dreams of exploration beyond the Earth, into space, and turning them into a chapter of human history. Practical steps have been taken. The consequences of these first steps are being considered all over the world, with a view to what we should accomplish in space, in the new millennium, especially in Europe. Europe has been the site of great philosophers, invention and technological development. Just as the debate continues in Russia, the US, China and Japan, Europeans should be ready to take part in the same progression, as humans decide how and at what pace we must move into space, to establish persistence of the human species beyond the fragile starting point, Earth.

Within the last 50 years Yuri Gagarin orbited the Earth for the first time, Neil Armstrong placed the first human footprints on the Moon, *Pioneer* spacecraft broke free of the solar system, *Viking* spacecraft soft-landed on Mars, and the *Voyager* and *Galileo* spacecraft have served as our proxies in the outer solar system. Put in this context, human exploration appears to be continuing, if mainly by robot proxies, at a respectable pace. However, to some impatient individuals, young hopeful space explorers, this rate of progress seems excruciatingly slow. When I was planning a career in 1981, NASA was hoping to launch a Space Shuttle every fortnight. As a graduate student, it did not seem at all unreasonable to me to plan a career to be in orbit by 1984, on the Moon by 1994, and on to Mars by 2004. This was the dream I penciled out as a plan, to set before myself. In hindsight, reality turned out to coincide with my rather optimistic plan in some significant ways, but differ markedly in others!

So what happened to redirect and distract space exploration over the last 25 years? Profound world events, including rapprochement between counter-posed political ideologies, the spread of democracy, industrial globalisation and economy, Earth science, pollution, and increased food production and improved distribution have at one time or another been on the minds of both politicians and electorates. When these factors coincided in some way with the conduct of space

exploration, then fortunate cosmonauts and astronauts became enabled to live and work in space, sometimes together, and sometimes
for many months. It is of world significance when superpowers, whose
nuclear arsenals threaten humanity's very existence, cooperate in
space. The long-lived Russian *Mir* space station and the partnership of
nations which make up the *International Space Station* may well have
been better first steps for the citizens of the Earth than any single nation planting its flag too soon on Mars, without broad global support.
Reality has turned out to be quite different from what I had hoped for,
when I was a young, and probably naive, aspiring space explorer. The
world events that took the place of those dreams are arguably more
profound historically for the majority of the world's population. This
makes me optimistic, both for the continued betterment of humanity,
and its future survival by inhabiting places other than just the Earth.

In fact, as the pace of human space-faring develops, ventures
in space will provide ever-increasing choices for us to consider, with
quite unpredictable consequences. This development will take place
no matter whether any one country or multinational consortium has
the determination to opt in or out of this adventure. No matter what
central or global planning is agreed upon for implementation, the
dreams and actions of individuals are already beginning to shape and
influence space development in haphazard and unpredictable ways.
A few years ago, nobody would have predicted publicly that enterprises of the former Soviet Union would be foremost in providing a
large fraction of space transportation costs by means of space tourism.
Now that space is no longer the single purview of governments, but
can be reached within the means of individual citizens, it will become
increasingly difficult for governments and nations to chart a slow or
restrained course for space development. A growing and diversifying
space economy, involving human settlement on the Moon and Mars
and the utilisation of all the solar system's resources, will be the result
of innumerable ventures only now being dared.

In Space the Final Frontier? If we are brave enough to tackle
refreshing ideas, which for the present are the stuff of dreams, science

fiction, and imagination, we find ourselves liberated. For indeed, it is our motivation to discover new science that benefits, inspired by the consideration of aliens, interstellar travel, and incredible changes to the current laws of physics. When we think of the possibility of new laws, new physics, new worlds, new partners, and new horizons – then we should hedge our bets. Space is only the Next Frontier.

C. Michael Foale
First British–American astronaut

Preface

Space, The Final Frontier is the *leitmotiv*, the theme phrase, of a very popular science fiction series on TV. *Star Trek* has, over the years spawned dozens of clubs whose members – *trekkies* – have developed a quite uncommonly optimistic vision of life and technology. If a *trekkie* were asked the question which constitutes the title of this book, the unhesitating answer would be 'Yes!' He or she would, in fact, be shocked that anyone could add a question mark to a statement believed to be a self-evident truth, even an article of faith.

Thousands of other people in industries, research centres and universities, who work on projects in some way related to the exploration or commercial exploitation of space, would also give an essentially positive answer to our title question. But their answers would be far less starry-eyed, far less enthusiastic. Many of them regard space an extension of their lab, a place where research can be performed in a better way. For others space is an ideal place for new equipment for improving Earth-based services. Some of them may even think that dealing with the long-term strategic goals of humans in space is a pointless exercise, an impractical hybrid of science fiction and real life.

Beyond this space-related professional world, a good proportion of public opinion is ready to enthuse about spectacular space enterprises, as they would about some sport exploits. However, they soon start to question whether it is worth our while to invest such large sums in something whose goals and uses they do not clearly understand.

Many others, on the other hand, would react quite differently, with answers ranging from a sceptical 'maybe' to a plain and annoyed 'no'. Reasons for rather negative answers may be their general anti-technological feelings or political, moral or even religious views. Some of them see space exploration as part of a process which is forcing humanity to lose sight of its very nature and values, pushing it toward a future in which technology replaces human values. If not stopped in time, that route could lead to the destruction of the ecosystem on which human life itself depends.

Other sceptics see space exploration as a toy for a privileged part of humanity. After seizing almost all of the planet's resources, they are now wasting them on costly and useless technologies. An even more cynical view is that space research is an instrument of the military/industrial/political establishment, always on the lookout for new ways to increase its wealth and power.

In reality, the greatest slice of public opinion holds an inter-mediate position between that of the space enthusiasts and of those who, for various reasons, are opposed to space research. This major-ity is excited about new images of the distant Universe and uses the benefits of space technology, such as improved long-distance telecom-munications or more-accurate weather forecasts. But it is essentially indifferent to space exploration, and unwilling to allocate more public funds to the space agencies around the world.

Forty years after the official beginning of the space age, and about a century after the first pioneering studies based on sound scientific knowledge, two basic attitudes among those who advocate human expansion in space are consolidating. Both of them justify the need for humankind to develop a presence beyond the Earth. Both assert that space *is* the true frontier for humankind, which must gradually be pushed out into the Universe surrounding us, with all the overtones which the word 'frontier' evokes.

The first attitude can be summarised in one statement: although humankind evolved on Earth, its destiny is to go beyond its home planet. It is the so-called *space imperative* for human beings to leave the surface of the Earth and spread into the Universe. Konstantin Tsiolkovsky synthesised this view in the most direct way by noting that, 'if it is true that the Earth is the cradle of mankind, it is also true that everybody must sooner or later leave the cradle in order to face life'. The history of humankind on planet Earth is just a sort of prologue for the real history, which has yet to begin – the history of humankind colonising the Universe.

This attitude can be traced back to Russian Cosmism, which flourished at the end of the nineteenth and at the beginning of the twentieth centuries, and whose main exponents were the philosopher Fedorov and Tsiolkovsky himself. Cosmism is said to have influenced many leading intellectuals of its time, including Dostoevsky and Tolstoy.[1] It preached the unity of humanity (or, more precisely, of the many communities of intelligent beings Tsiolkovsky believed to have evolved on different star systems) and the Universe, and affirmed the cosmic destiny of humankind.

This vision has many different interpretations nowadays, from the more limited one, that the human species will forever be confined to the solar system, to those predicting expansion on a much larger scale. The most straightforward reason for the space imperative is the simple need to survive; it is not wise to 'put all our eggs in one basket' on a planet which is subjected to cataclysmic events of various kinds[2] as it follows the fate of our solar system. Other reasons are materialistic, historical or religious. In the latter case, the human species is seen as the only intelligent being in the Universe, the agent whose duty it is to humanise Creation, or as one of many intelligent species whose purpose is to fulfil the vision of a cosmic community. A destructive criticism of this, which accuses it of demonic arrogance, can be found in *Out of the Silent Planet*[3], written two decades before travelling through space became a real possibility.

A more pragmatic vision of the space imperative has been introduced since then. Humanity must colonise space to overcome the

[1] V. Lytkin, B. Finney, L. Alepko, K. Tsiolkovsky, Russian cosmism and extraterrestrial intelligence, *Q.J.R. Astron. Soc.*, vol. 36, pp. 369–376, 1995, and B. Finney in *Keys to Space: An Interdisciplinary Approach to Space Studies*, A. Houston and M. Rycroft (editors), McGraw-Hill, New York, 1999.
[2] British National Space Centre, *Report of the Task Force on Potentially Hazardous Near Earth Objects*, Her Majesty's Stationery Office, London, 2000.
[3] C.S. Lewis, *Out of the Silent Planet*, John Lane, London, 1938.

intrinsic limits to resource availability on Earth, limits which are slowing down and which will, in the future, jeopardise its further development. The conclusions reached in the famous *The Limits to Growth* report, drawn up for the Club of Rome[4] at the end of the 1960s, are extremely clear. They put an end to the dream of humankind's unlimited growth on planet Earth. Although this is not the place to discuss the polemics raised by the report or the interpretation of its results, it clearly only applies to the situation in which the human species exists on a single planet and has access only to the resources available on the Earth.

This second basic attitude to our expansion into space is the so-called *space option*, i.e. the possibility of importing resources from space, whose production would not contribute to the pollution of our biosphere, while at the same time sending a significant number of human beings from our planet to work in space. In this way the well-being enjoyed by some could be extended to all. Such an approach is controversial, however, as there are doubts that the space option could be implemented either at an acceptable cost or on a large enough scale to make a substantial impact on the global development of humankind. While the space imperative deals with long-term perspectives, the space option offers a solution to a crisis which might be imminent, at least if we believe the projections on which *The Limits to Growth* is based. Should we embark upon the space option without delay, before the lack of available resources makes it less effective, or altogether impossible?

Whatever we decide to do, we must consider that our expansion into space will change us deeply. *Homo sapiens* has changed very little, from a physical point of view, since the first appearance on this world, but has advanced mainly through an accumulation of knowledge and technological know-how. Living in space in environments which are so different from conditions on Earth will cause

---

[4] D.H. Meadows, D.L. Meadows, J. Randers, W.W. Behrens, *The Limits to Growth: Report for the Club of Rome's Project on the Predicament of Mankind*, Potomac Associates, New York, 1972.

adaptation processes to occur and might change individuals in a few generations.[5] Those who live in a space colony would only be able to return to live permanently on the Earth after a long and painful re-adaptation. Whereas the physical consequences can be predicted, psychological issues could be equally important. What will be the re-action of a human being born and brought up in an artificial space environment when he or she goes out into the open on the surface of the Earth?

With time the human species could differentiate into many different species, attuned to various environments. What will a society be like that is not only multi-racial but, to coin a new word, *multi-speciated*? And what will happen if, during this expansion, human beings come across alien lifeforms, perhaps intelligent ones? All this leads to problems that are more moral than scientific or technological.

Is it morally acceptable to *terraform* a planet,[6] i.e. to modify it so as to make it possible for human beings to live there? What if, by doing so, the lifeforms which developed on its surface become extinct? And what if that life includes intelligent species? And who will give the patent of intelligence to a species? Who will decide to what extent it must be respected and protected?

That science and technology today are no longer morally neutral is obvious, and bioethics is a good example. It may be no more impossible to terraform a planet in the slightly more distant future than, in the near future, to clone a man or woman or than it is today for one woman to deliver another woman's baby. Bioethics teaches us that waiting for technology to make some things possible before we start a discussion about them leads to conflicts and uncertainties. We need to consider future possibilities now, so as not to be compelled to make decisions in haste, under the pressure of events and unavoidable interests.

---

[5] F. White, *The Overview Effect: Space Exploration and Human Evolution*, The American Institute of Aeronautics and Astronautics, Reston, Virginia, 1998.
[6] M.J. Fogg, *Terraforming: Engineering Planetary Environments*, Society of Automotive Engineers, Warrendale, Pennsylvania, 1995.

The main purpose of *Space, The Final Frontier?* is to illustrate the possibilities which first the exploration and then the exploitation of space will open up for humankind, to discuss the factors which may push it along this road, and to look to the future. The matter will not be dealt with in a chronological way, even though we shall be following a future – and somewhat hypothetical – history. It will rather be developed in concentric circles, like ripples on the surface of a pond after a stone has been dropped in, starting from where we are now. By doing this, a sort of future chronology will sometimes be followed, as it is likely that nearer targets will be reached earlier than more distant ones. But this is not necessarily true; the first interstellar probes will probably be launched well before the environs of Mars are modified by human beings.

This book does not claim to answer all the possible questions posed by its title. However, its purpose is to supply the reader with information which he or she can use to put space in perspective, and to approach space issues in a more rational way, hopefully devoid of prejudice and biased views. Obviously, the authors have their own opinions and biases about many of the topics discussed, and they do not claim to be able to supply an absolutely objective viewpoint. But they attempt to report facts and opinions correctly.

The aim is not try to make technological forecasts: forecasting future technologies is often pointless, as really innovative technologies are unpredictable. Who, for example, could have forecast the current information revolution back in the 1930s? Moreover, the proposals and recommendations for future goals in space exploration, or the priorities in the development of new technologies, put forward by various *ad hoc* commissions and committees (of space agencies, of the International Academy of Astronautics or of national parliaments) continually seem to shift their focus. To try to stick to such scenarios would be impossible, and would make this book obsolete even before it was printed. And new concepts are being introduced all the time.

Also there is no attempt to predict what specific actions might be taken in future, nor when the various steps of human expansion in space may occur. As an example, when speaking of the colonisation of Mars, we do not consider that Mars will be colonised within a certain timeframe, or with given technologies. Rather, we discuss the implications for humankind, the difficulties, the dangers and the payoffs.

Nor is the aim of the book to cover the history of human expansion into space and of the technologies which made that possible. An outline of space history and of the technologies needed to understand future perspectives cannot be avoided, but they are kept short.

An outline of the material presented is as follows.

The first chapter gives a short account not only of present activities in space but also of the importance of space activities in the recent past and nowadays. An attempt is made to illustrate the different opinions held about some controversial aspects of space exploration, such as the causes of various crises, the importance of the Space Shuttle, and the value of having humans in space.

The second and the third chapters are devoted to activities in space near the Earth, mainly the rocket propulsion required to launch satellites into orbit around the Earth, and space stations.[7] Ways to explore the various planets of the solar system in the relatively near future are dealt with in Chapter 4, which gives an account of the robotic devices required to reach its outer limits.

The human exploration and colonisation of the solar system are studied in the next three chapters. These deal, respectively, with our return to the Moon, the colonisation of Mars and the exploration of the other planets, their satellites and the asteroids, in the more distant future.

The last part of *Space, The Final Frontier?* moves from the arena of science fact into the realm of science fiction, of imaginative ideas

---

[7] Technical terms are defined and explained in M. Williamson, *The Cambridge Dictionary of Space Technology*, Cambridge University Press, Cambridge, 2001.

based upon scientific principles. The automatic and then the manned exploration of interstellar space which may become feasible in the long term future is the subject of Chapter 8. Attention is focused on radically different methods of propulsion through space. The possibility that humans may eventually leave the solar system at all is a very controversial subject.

The ninth chapter deals with the fascinating theme of the search for life, and for intelligent life in particular, away from our planet. Such a subject, on which many books[8] have been written, will not be studied in detail, but only in relation to those aspects which are relevant to the exploration of space. The aim is mainly to discuss the possibility that humans, during their expansion into space, will encounter living beings, possibly intelligent ones, which differ from *Homo sapiens*.

The concluding chapter looks further into the future, trying to imagine the consequences of those revolutionary technologies which science fiction has popularised, and the new, wider perspectives which will be gained.

Finally, there are a few appendices for readers who feel the need to dig deeper into the scientific and technological aspects of space travel (the distances to be crossed when travelling into space and the related units, the dynamics of space navigation and the principles of propulsion devices). They are introductions to subjects which require a far deeper study, plus advanced mathematical and physical knowledge. The reader who is more interested in the human aspects of space exploration than in the technological issues can skip these with advantage.

The authors are most appreciative of the countless conversations with their many colleagues and students during their careers in

[8] S. Dick, *Life on Other Worlds: the 20-th Century Extraterrestrial Life Debate*, Cambridge University Press, Cambridge, 1998, B. Jakosky, *The Search for Life on Other Planets*, Cambridge University Press, Cambridge, 1998, P.D. Ward and D. Brownlee, *Rare Earth: Why Complex Life is Uncommon in the Universe*, Springer-Verlag, Berlin, 2000.

space which have formed the basis for this book. They express their gratitude to their respective wives, Franca and Mary, who have devoted much time to reading and commenting constructively on the text. Finally, they thank the reviewers for their advice; they are most grateful to many people at Cambridge University Press, especially Jacqueline Garget and Simon Mitton, for their efforts during the preparation of their book for publication.

Giancarlo Genta
Michael Rycroft
February 2002

I   Space today

The space age has had more than its fair share of ups and downs in its brief, 45-year history. Three key questions are:

- Why did our vision of space exploration fade so rapidly after the *Apollo* programme to the Moon?
- Should space activities be regarded as a rather special way of making money, or of viewing the Earth and the Universe beyond?
- Do astronauts and cosmonauts do a much better job than robots in space, and so justify the much greater expense involved?

People with different backgrounds will give different answers to these questions.

A DRAMATIC BEGINNING

Conventionally the space age began on October 4, 1957, when the first artificial satellite of the Earth, *Sputnik 1*, a small spherical object with a mass of 83.6 kg, was launched by the Soviet Union. Up to that day it seemed that only a relatively few people were interested in whether man could learn how to travel in space, while most people were either disinterested or doubtful about its actual feasibility. Pioneers such as Goddard, Tsiolkovsky, Oberth and von Braun anticipated space travel. They worked hard to make their dreams come true, often encountering scepticism and criticism on the way.

In 1926 a British scientist, A.W. Bickerton, wrote:

This foolish idea of shooting at the moon is an example of the absurd length to which vicious specialisation will carry scientists.

> To escape Earth's gravitation a projectile needs a velocity of
> 7 miles per second. The thermal energy at this speed is
> 15,180 calories [per gram]. Hence the proposition appears to be
> basically impossible.

Space travel seemed to belong more to fiction, particularly to science fiction, than to science or technology. Many men of letters were much interested in it: as an example, the Italian writer Carlo Emilio Gadda wrote in 1952 about the exploration of the Moon, Venus and Mars, defining these celestial bodies as the *New Indies*, a definition which implies not only exploration but also colonisation.

On October 4, 1957, the whole picture suddenly changed. What had previously been considered with scepticism or regarded as impossible turned out to be within the range of human capabilities, and new hopes emerged. In the Western World, however, these were accompanied by the fear of losing the technological supremacy on which its very survival seemed to rest – that Cold War atmosphere might, at any moment, degenerate into actual war. These feelings became stronger when, on November 3, 1957, the Soviet Union repeated its success, launching an even larger satellite. It had a mass of 508 kg and carried an animal, a dog named Laika. The demonstration that life was possible aboard a satellite in orbit around the Earth was there for all the world to see.

It was such a shock that there were some who could not believe these facts, arguing that it was all a form of propaganda. Within about three years notable achievements were recorded by the Soviet Union and by the USA – the first Moon probe (*Luna 2*, 1959), the first images of the other side of the Moon (*Luna 3*, 1959), the first weather satellite (*Tiros*, 1960), the first probe to Venus (*Venera 1*, 1961, and *Mariner 2*, 1962), and the first man to make an orbital flight (Yuri Gagarin, on *Vostok 1*, April 12, 1961, Figure 1.1).

President John Kennedy, in his famous speech of May 25, 1961, declared formally that the United States would land a man on the surface of the Moon and return him safely to Earth before the end

Figure 1.1. The first two astronauts: on the left Yuri Gagarin, who travelled once around the Earth in the spacecraft *Vostok 1* on April 12, 1961; on the right Alan Shepard, who performed a suborbital flight on the spacecraft *Mercury Freedom 7* on May 5, 1961 (the picture refers to the *Apollo 14* mission, during which Shepard walked on the Moon).

of the decade. This goal was spectacularly achieved.[1] Such an ambitious programme had never before been attempted and has never again been matched. The old NACA (National Advisory Committee for Aeronautics) was transformed[2] in 1958 into NASA (National Aeronautics and Space Administration), with powers and funding far greater than before.

Their gargantuan efforts enabled the United States to reduce the large lead which the Soviet Union had in the space race.[3] Not only

[1] W.E. Burrows, *This New Ocean: the Story of the First Space Age*, Modern Library, New York, 1999.

[2] R.E. Bilsten, *Orders of Magnitude: A History of NACA and NASA, 1915–1990*, NASA, Washington, DC, 1989, including works of art from NASA's collection; see R.D. Laurins and B. Ulrich, *NASA & the Exploration of Space*, Stewart, Tabori and Chang, New York, 1998.

[3] M. Collins, *Space Race: The US–USSR Competition to Reach the Moon*, Pomegranate, San Francisco, California, 1999.

were astronauts sent to the Moon, but also a number of other goals, such as new rocket launch vehicles, satellites of all types and robotic exploration of the whole solar system, were met in the 1960s and into the early 1970s.

UNFULFILLED PROMISES

Everything seemed possible in the enthusiastic atmosphere of that time, even if a few voices were starting to object to human expansion into space as summarised in the Preface. But who could cast doubts on the technical feasibility if such an unbelievable enterprise as landing a man on the Moon and bringing him safely back to Earth could be performed in just eight years? A few might have said that it was not worth going to Mars or reaching any of the other proposed goals. But they did not say that such goals were beyond the reach of humans, and in a short time too. Several ambitious projects started to take shape.

As a prerequisite a fleet of 'aeroplanes' capable of reaching low Earth orbit was needed. Many design projects were carried out and it was predicted that such flying machines could be operational by the end of the 1970s. This schedule was essentially met, but with severe limitations. The Space Shuttle (Figure 1.2), flight tested in 1981, is a reusable launcher which lands like an aeroplane. However, severe compromises had to be made on its complete reusability – it has two external solid fuel boosters which are recovered from the ocean and a large non-reusable external fuel tank. The Space Shuttle's operational costs are enormous and its performance characteristics not as good as had been hoped for. That routine access to space, of which its promoters were dreaming, was not realised.

A second huge project was that for a space station. The designs of the 1950s anticipated a very large space station, possibly like a wheel (Figure 1.3), slowly rotating about its axis to obviate the lack of gravity with the centrifugal acceleration due to its rotation. It was anticipated that, by the end of the 1970s, several crew members would live, more or less permanently, in space. They would work in the various sections

Figure 1.2. Landing of flight STS 67, the Space Shuttle *Endeavour*, on March 18, 1995 (NASA photo).

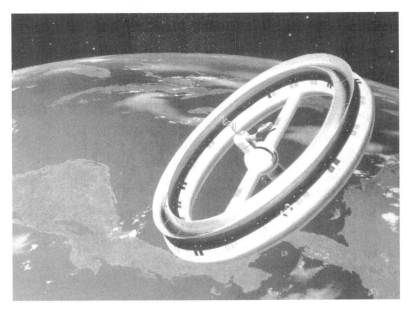

Figure 1.3. A space station in the shape of a wheel, as it was dreamt of by space pioneers and such as appeared in movies like *2001, A Space Odyssey*.

of the space station including scientific laboratories, factories, a spaceport and even a hotel. These predictions turned out to be wrong, and
not only as far as the timing is concerned. The construction of the
International Space Station started at the end of 1998; large as it is,
it will be smaller and simpler than the space station originally conceived. It will be a step forward in the direction of the colonisation of
space. Thanks to its modularity, it may be enlarged following a possible future redefinition of needs to be a staging post in space. However, it will always fall far short of the expectations of the 1960s and
1970s.

When planning journeys to the Moon there was the idea that humans would go there to live. Many other missions, longer and more
complex, could have followed the first *Apollo* landings, until a permanent outpost was built, perhaps in the 1990s.

A short summary of the predictions of the 1960s cannot avoid including a human landing on Mars, which was assumed to have taken
place, without doubt, in the early 1980s. Optimistic forecasts were
made in the 1960s, when all seemed possible and when the time between any one achievement and the next, more ambitious one was
incredibly short. But the entire history of space exploration is packed
with unfulfilled promises and erroneous forecasts. At the end of 1988,
in one of the darkest periods of space exploration, and following the
recommendations of a Working Group chaired by the astronaut Sally
Ride, after the *Challenger* disaster, NASA recommended, among other
goals, the following:

- a permanent base on the Moon,
- a manned expedition to Phobos, one of the satellites of Mars, in 2003,
- a manned expedition to Mars, in 2007.

Needless to say all these goals are very far from being achieved.

Often a mission with very ambitious goals is proposed and the
budget initially put forward seems to be adequate. Then, when the
initial study (phase A study, in aerospace jargon) proceeds, the costs

rise and the funding is relatively reduced, forcing the mission goals to be less ambitious. With the subsequent design (phase B), development (i.e. construction, phase C) and implementation (phase D) stages, further reductions of the objectives often take place in such a way that the actual mission is but a pale imitation of the initial one.

Most important missions since the mid 1970s have been either downsized or cancelled altogether. Even successful missions, such as the US *Galileo* probe to the planet Jupiter, had been rethought, redesigned, delayed and even risked being cancelled. Another example is the *Rosetta* mission, an initial objective of which included the retrieval of samples from a comet: it has been simplified, and now will just land on the nucleus of a comet.

The consequences of this failure to meet expectations are great. The general public has lost much of its initial interest in space exploration, and the effects on many specialists have been devastating. To cancel a space programme, or to make budgetary cuts to a project which has already started, often means reducing the number of personnel employed in research centres or in divisions of the space industry. Hundreds of space specialists have been dismissed or moved to less-creative jobs within the same industries working for space agencies. Apart from these human problems, it is a waste of human resources, both of highly qualified individuals and of the huge investment of time and effort in creating effective teams. However, some positive aspects can be found even in this process, as some specialists who found new jobs in other sectors brought to the latter the valuable experiences which they acquired in the space field; performing technology transfer in this way is, nevertheless, extremely inefficient.

There is a worse, and more subtle, problem. Those who have worked on programmes which are cancelled are demotivated, and may develop an attitude towards their work which is bureaucratic. If it is likely that a mission will never 'fly' in space, what matters most is not the mission itself but the number of formal duties, meetings, progress reports, assessments and other paperwork by which the financing agency justifies payment for the work performed.

All those who work on a space programme need to be sure that the programme will be completed, even if this means that the choice of missions to be financed must be made with more severe – and realistic – criteria.

CRISIS OF GROWTH?

The crisis of space activities started exactly when space was experiencing its greatest triumph. When the *Apollo* programme was at the point of taking a man to the Moon, the political situation was very different from that at the beginning of the 1960s. The Vietnam war was attracting the attention of public opinion and there were strong protests in many countries, especially the United States, against the American administration. The enthusiasm of the Kennedy era for the 'new frontier' was being substituted by duller national politics and the credibility of US and other institutions was decreasing. Anti-technological movements, with their ecological aspects but also with their sheer irrationalism, were gaining ground. The primacy of politics, preached by the movements of 1968, put all other activities in a shadow of suspicion or contempt.

The first two landings on the Moon (*Apollo 11* and *Apollo 12*) attracted considerable attention for a short time and brought the earlier enthusiasm to a climax (Figure 1.4). But with *Apollo 13* nearly ending in tragedy, criticisms were voiced and requests to cancel the *Apollo* programme gained momentum. And this was so notwithstanding the fact that *Apollo 13* dramatically demonstrated that human beings were able to work in space, reacting to unpredictable events and showing a surprising ability to improvise in order to master the worst of situations.

Only the success of the *Apollo 14* mission could save the programme, which had already been downsized – it enabled the last three *Apollo* missions, *15*, *16* (Figure 1.5) and *17* to proceed. And even *Apollo 14* came very close to failure; the docking of the command module to the lunar excursion module, a manoeuvre which was essential to transfer the crew for the Moon landing, almost failed. It was

Figure 1.4. Picture of Edwin 'Buzz' Aldrin, lunar module pilot, on the Moon taken by the mission commander Neil Armstrong; he had just deployed some scientific instruments, visible in the foreground (NASA photo).

attempted five times without success; it was only when the pilot of the command module tried to force the secondary locking device by ramming the two spaceships together at a speed greater than anticipated that the docking was successful. Later, it was discovered that the mechanism had been jammed by an ice crystal. In retrospect it is amazing to contemplate that the future of a multi-billion dollar programme and four of the most important *Apollo* missions had been jeopardised by a relatively simple mechanical device and by an insignificant ice crystal. Lunar missions ended with *Apollo 17* (Figure 1.6). Many thought that all the goals of the entire space adventure had been reached, but in a somewhat disappointing way. Some thought that the lunar missions were only a 'space show', an effort to impress public opinion rather than to attain scientific goals or to open the space frontier to further projects with crews. The result of this approach could

Figure 1.5. The *Apollo 16* Mission; the Lunar Excursion Module is on the surface of the Moon and an astronaut is getting the Lunar Rover ready for travelling (NASA photo).

only be a disappointment, with no continuity in the detailed exploration of the Moon to lead on to the construction of lunar outposts.

Other people noted that the disappointment was unavoidable. The decision to send men to the Moon instead of robotic probes, which was taken against the advice of many members of the space community, was also based on the consideration that it would be easier to obtain the required funding for a mission making a very high impact on public opinion.[4]

But this is highly questionable. The *Apollo* programme was far more than a series of 'flag and footprint' missions; the astronauts carried out top-class scientific research on the Moon, and the results of

---

[4] A detailed account of the discussions which accompanied the *Apollo* programme and of the scientific, political and military milieu in which they took place can be found in the form of the novel *Space*, by J.A. Michener, Corgi Books, London, 1986. A factual account is given in B. McNamara, *Into the Final Frontier: The Human Exploration of Space*, Harcourt, Orlando, Florida, 2001.

Figure 1.6. An *Apollo 17* astronaut performing geological studies during one of the long-range reconnaissance activities using the Lunar Rover (NASA photo).

the *Apollo* missions shed new light on the origin of the Moon and of the solar system. The ease with which the astronauts learned how to move and how to work in such an unusual environment showed that colonising extraterrestrial bodies is not as difficult as many predicted then and still think now. Advances in the engineering sciences brought about by the *Apollo* programme were outstanding; in retrospect we can say that they changed our lives. Every time we walk on snow, use a computer or travel by plane we use something developed for the Moon adventure or designed using methods introduced for it. And this trend is continuing. Soon we will drive non-polluting cars powered by fuel cells which would never have been developed without the *Apollo* programme.

The *Apollo* programme was also good for business. Costing some US $70 billion (in present-day currency terms) its revenues, in royalties and know how, were several times that. Some economists consider that the unprecedentedly long expansion phase of the American economy which is still going on was founded on the technological effort following John Kennedy's commitment to land a man on the Moon.

The decision to terminate the *Apollo* programme was a political decision taken by Richard Nixon under the influence of current pressures and for short-term economic convenience.[5]

Space Shuttle: instrument of progress, or hindrance?
In fact, NASA was concentrating on the post-Apollo programme, with the Space Shuttle as one of its focal points, but without any general agreement among the specialists as to where space efforts should be focused. On the one hand, there was the widespread opinion that only a launch machine which could dramatically reduce the cost of launching satellites could really open the gateway to space. On the other hand, some thought that concentrating the shrinking resources into a launch vehicle would reduce the funds available for scientific missions. So it was a quarrel between engineers and scientists, between those who wanted to put the emphasis on technology and those who wished to stress science. There was a further disagreement between those who thought that the role of humans in space was essential – the Space Shuttle being a vehicle to carry humans beyond the Earth's atmosphere – and those who promoted the robotic exploration of space. The Space Shuttle was even seen as a plot by the industrial/military complex to use public funds to build a machine whose importance was mainly military.

A similar argument, which is in fact still going on, was later initiated by the decision to build a space station. It is likely that such

[5] See H.E. McCurdy, *Inside NASA: High Technology and Organizational Change in the U.S. Space Program*, The Johns Hopkins University Press, Baltimore, Maryland, 1993, and R.D. Launius, *NASA: A History of the U.S. Civil Space Program*, Krieger Publishing Company, Malabar, Florida, 1994.

debates will follow future decisions regarding the construction of a lunar base, the landing of men on Mars, or future ambitious projects in space.

Both sides had good reasons to support their viewpoints, and the main problem was basically one of politics and economics. In a situation of shrinking budgets it was difficult to cope with the operating costs of the Space Shuttle, which were found to be much higher than expected. And now the Space Shuttle is ageing. Technology has made much progress since the Space Shuttle was designed and built, above all in electronics and computer science, but also in materials science and design techniques. The current research programmes, such as NASA's X-33 programme[6] to build a second generation shuttle, must proceed rapidly, since the pressure to reduce launch costs is increasing. The current NASA maxim of 'faster, better, cheaper' can be applied, to a certain extent, to launch vehicles, satellites and spacecraft.

It is also very important not to entrust all space activities to a single type of launcher, but to have some elements of competition between launch manufacturers. The very existence of the Space Shuttle and the necessity to justify its enormous and escalating costs led to the construction of large non-reusable rockets being stopped in the United States. A launch system based on a single class of vehicles is extremely vulnerable to a 'single-point failure'. Only four Space Shuttles were built, and such a failure occurred with the *Challenger* tragedy of January 28, 1986. The grounding of all Space Shuttles after the *Challenger* disaster paralysed American space activities for several years. The lesson to be learned is that the different needs of the various space missions planned require some choice between several different launch vehicles.

The fate of the *Galileo* probe, to study Jupiter and its satellites, is a good example of the problems associated with the use of the Space Shuttle. Initially, the launch of that probe was scheduled for 1981. It should have left Earth orbit aiming directly at Jupiter under the thrust of a chemical rocket, after having been put into low Earth orbit

---

[6] The X-33 programme is now cancelled, but other programmes are replacing it.

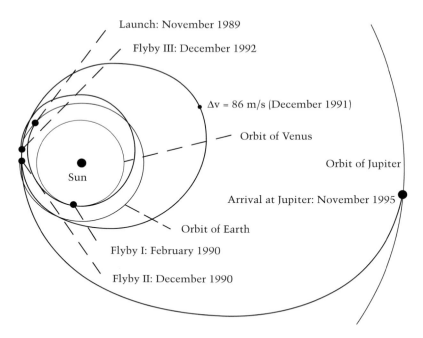

Figure 1.7. Trajectory of the *Galileo* probe, with the flybys shown and also the trajectory correction, a boost giving a velocity increment Δv of 86 m/s (see Appendix B).

by the Space Shuttle. The *Challenger* disaster delayed the launch for some years and new safety regulations prevented the liquid propellant Centaur rocket from being carried aboard the Space Shuttle. Then the Centaur programme was cancelled and it seemed that the *Galileo* mission would have to be abandoned. Considering the impossibility of a direct launch, a complete redesign of the mission was undertaken. The trajectory eventually chosen was for a launch toward Venus, with a close flyby to use 'gravity assist' due to that planet[7] to obtain a first increase of speed.

The very complicated trajectory later included two flybys of Earth to aim the spacecraft towards Jupiter (Figure 1.7). The total travel time was greatly increased, from 2 to 6 years, and so were the chances

---

[7] A flyby with gravity assist is a manoeuvre in which a spacecraft exploits the gravitational field of a planet to change its trajectory and to increase its speed *en route* to its destination (see Appendix B).

of an accidental malfunction or of an error in the trajectory correction. A secondary antenna had to be fitted to the probe, to allow the primary one (which could not withstand the heat from the Sun during the initial part of the flight in the inner solar system, near the orbit of Venus) to be folded initially. All this increased the complexity of the mission considerably.

Galileo was eventually launched on October 18, 1989 by the Space Shuttle *Atlantis*. The trajectory proved to be feasible and, after passing close to Venus in February 1990, to the Earth in December 1990 (see Figure 1.8), to the asteroid Gaspra in October 1991 (Figure 1.9) and again to the Earth in December 1992, *Galileo* began

Figure 1.8. South America and Antarctica as observed by the *Galileo* probe during a flyby of the Earth (NASA photo).

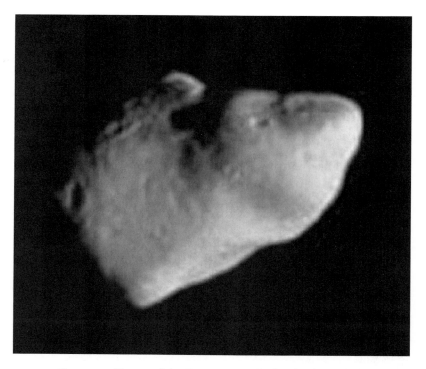

Figure 1.9. Picture of the Gaspra asteroid taken by the *Galileo* probe from a distance of about 16,000 km on October 29, 1991. Gaspra is 17 km long and 10 km wide (NASA photo).

to travel towards Jupiter. When the main antenna failed to open, this malfunction seemed likely to lead to the total loss of the mission. But with the probe homing in on Jupiter and its main antenna inoperative, the programming of the on-board computer was changed to include recently developed image compression techniques. In this way images were broadcast, although at a far slower rate than planned originally, using the secondary antenna (Figure 1.10).

The mission was finally a success; the atmospheric entry module was released in December 1995, while the orbiter continued its observational work around the planet. The arrival at Jupiter had been delayed by 12 years, more than the time which elapsed between President Kennedy's speech and the actual landing on the Moon! On

Figure 1.10. Image of Jupiter's moon Io taken by the *Galileo* orbiter during its ninth orbit around Jupiter. An enormous volcanic plume is clearly visible (NASA photo).

the positive side, however, was further confirmation that the gravitational fields of planetary bodies can be used to change a space probe's direction and speed just as old-time sailors exploited the winds and ocean currents. It also demonstrated that a mission which seems to be doomed to failure can be saved.

Space bureaucracy

In the 1960s NASA had an almost perfect reputation and considerable popularity; it looked as if nothing was impossible. The Americans were achieving success after success – they were regaining the technological supremacy which, at the end of the 1950s, they seemed to have lost.

Then came budget cuts, and downsizing become the 'in' word. Many of those involved in the space triumphs became disillusioned; some looked for more interesting jobs, others were made redundant or retired, and some died. Among the latter was Werner von Braun, who died in 1977. As often happens, the decline of enthusiasm and of the *esprit de corps* led many to concentrate on the daily routine and the space agency became more and more like any other government bureaucracy. T.R. McDonough[8] notes that the rules of the civil service – under which most NASA centres operate – make it very difficult to sack the incompetent or lazy, and that the 'rocket scientists' and engineers were replaced by administrators more accustomed to 'flying a desk than a rocket, more adept at handling paperwork and office politics than nuts and bolts engineering'.

The Presidential Commission investigating the *Challenger* disaster included the Nobel laureate Richard Feynman, the astronaut Neil Armstrong and the test pilot Chuck Yeager. With the care typical of American Commissions at difficult times, they revealed the attitudes of complacency and overconfidence that led to that tragedy. The solid fuel booster blew up due to the failure of a small rubber seal between two of its segments. The examination of some boosters used in previous flights, which were recovered from the Atlantic Ocean,

[8] T.R. McDonough, *Space, The Next Twenty-five Years*, Wiley, New York, 1989.

had already suggested that possible problem. The engineers of the firm responsible for the boosters, Thiokol Corporation, had warned NASA not to launch the *Challenger* that day as the previous night's freezing temperatures could have damaged the seals. But those responsible for the launch continued, without even informing the doomed astronauts aboard of this. Whilst such a structural failure was the specific cause of this disaster, the actual cause was pressure exerted on NASA to keep up the schedule of frequent Space Shuttle launches, following the decision to launch all American satellites using this vehicle.

What applies to NASA bureaucracy could also hold for the European Space Agency (ESA), with added problems associated with the subtle rules for the division of the costs and benefits among its member states – namely the principle of *juste retour* – and of the overgrown and bureaucratic structures of the European Community.

Among the solutions to such problems is a reduction of the obstacles for private organisations to put payloads into orbit. Thus a privatised launch industry can be started; launch vehicles can not only be built, but also operated, applying the criteria of commercial enterprises. This could enable space agencies to concentrate on scientific missions and on those infrastructures which cannot be operated by private organisations but which, if correctly run, can be good investments.

A new, business-like approach will also cut costs. As Robert Zubrin notes in his book *Entering Space: Creating a Spacefaring Civilization*,[9] one of the causes of the very high costs of American space hardware can be traced back to the so-called 'cost-plus' system. Its aim is to prevent private enterprises from making too large a profit from a government contract. Instead of fixing a price in advance, as in private business, the companies supplying goods and services to the government agency must document all expenses and internal costs and then apply a limited profit (in the 10% range). In this way there

---

[9] Robert Zubrin, *Entering Space: Creating a Spacefaring Civilization*, Tarcher/Putnam, New York, 1999.

is an incentive to make every product cost as much as possible and, even worse, all investments on innovation are discouraged.

For the USA, the protectionist policy of using only American launchers for government-funded spacecraft has the effect of discouraging research aimed at reducing launch costs or improving the performance of rockets. A similar situation also affects Europe, with the additional fact that many aerospace companies are state-owned and, like most industries of this type, their efficiency is not very high.

Continuity of funding

In the *Bible* a well known parable of the New Testament asks the question: 'for which of you, desiring to build a tower doth not first sit down and count the cost whether he have wherewith to complete it?...'.[10] That is common sense, but it does not necessarily seem to be in agreement with the rules of modern society, where public funding is concerned. NASA must submit its annual budget to the Congress, who must approve the various line items and re-examine its projects every year; the other space agencies face the same, or similar, financial situations.

The problem stems from the fact that all space programmes, by their very nature, must last for several years. The tendency, at least for major space projects nowadays, is to go on for more than a decade. Nobody can guarantee that a programme, which has been duly approved and funded for a certain number of years, may not later be cancelled, perhaps when it is near to reaching its goals. Then the result is suddenly to nullify all the efforts made up to that point, with a tremendous waste of resources. Resources of all types, both human and material, will then be dissipated. The most striking aspect, a half-finished space vehicle rusting in a warehouse, is just the tip of the iceberg.

This does not mean that space activities must be removed from the control of the representatives of those who pay the bill. The choice of which space projects are to be funded and of their relative priorities

[10] St. Luke's gospel, chapter 14, verse 28, Revised Version.

is a political decision; that choice must be made by those to whom political affairs are entrusted. The problem is that, once the decision to fund a project has been taken, funding should ideally be guaranteed until the project comes to its natural end, that is after the observations made have been fully analysed and the results published. Of course, appropriate controls are needed on how the money is spent and on how the work is done. But it must be certain that, if the foundations of the tower in the parable are properly laid, the money to cope with escalating costs and to complete the roof will, when the time comes, be made available.

This implies that all concerned must make very serious commitments. In the first instance, the committees of the space agencies must choose from among the various proposed missions those which guarantee the best benefit/cost ratio and meet in the best ways possible the priorities as stated by the authorising bodies. Then, too, the aerospace industries must do their best by making realistic cost estimates, and preventing the costs from rising exponentially between the initial assessment phase and the final construction. Obviously, when a new project involves innovative technologies there are always some risks and financial uncertainties, but the due margins must be allowed with transparency and sincerity. Once the project has been started, it should be cancelled only in the event of a major, and unforeseen, problem, which renders the project no longer feasible or if its costs have soared in an unbearable way.

Political will
In this attempt to analyse the causes of the various crises which have hampered space activities since the middle of the 1970s, the fact that reasons of national pride are a motivating force must not be forgotten. Space was initially used as a spectacular stage on which to display the superiority of a political and ideological system, of a military power exceeding that of the opposing side. This was true for both sides which, during the Cold War, were confronting each other – the USA and the Soviet Union.

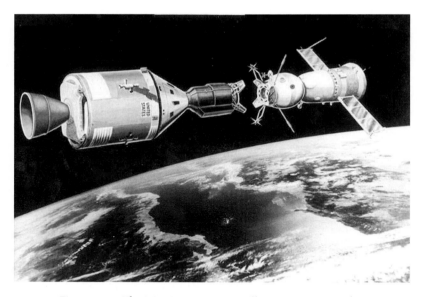

Figure 1.11. The joint USA–USSR *Apollo–Soyuz* mission of 1975. Artist's impression of the two spacecraft just before docking (NASA image).

Almost unlimited resources, material and human, could be thrown into the battle. After the landing on the Moon, changes were afoot; even though national rivalry and confrontation were still present, they stopped playing such an important role in the space race. The joint *Apollo–Soyuz* mission of 1975 (Figure 1.11) was pivotal in the switch from confrontation to co-operation, even with its inevitable ups and downs and with varying degrees of success. Today, with the changed international situation after the Cold War, it is necessary that the greatest space powers redouble their efforts to collaborate. The first phases of the construction of the *International Space Station* (Figure 1.12) have been carried out. There are several exciting missions to Mars, the red planet, in prospect, and the search is on for Earth-like planets orbiting nearby stars.

But international co-operation in itself is not the solution to all problems. There are instances when that causes such a large increase of the cost of the mission that the share of some participants

Figure 1.12. Artist's impression of the Space Shuttle docked at the *International Space Station* (NASA image).

is not much smaller than the cost of the whole mission if carried out by a single country. First, there are the technical difficulties of defining common standards and design practices for industries working with different traditions, experiences and even measuring units. An example is the various components of the *International Space Station*, which must mate perfectly in the difficult environment of low Earth orbit, but without any trial assembly having first being made on the ground. But the worst problems are of a political nature: sometimes the different partners insist on using their own system to perform a certain function, to the point that some subsystems are duplicated – formally to increase safety through redundancy, but actually to reach an agreement. Each country participating in the common enterprise tries to obtain a large part of the control infrastructure and, when the mission includes a crew, of the training facilities on its own territory. These ground facilities allow the paying public visibility of the space project. They produce employment – each politician involved would like to obtain one facility in his or her own constituency.

Each interface must be defined through long and difficult negotiations, with each player trying, for example, to ensure that his or her own standards are used, because he or she actually thinks that they are better, and do not necessitate any changes to usual practices. The delays associated with such negotiations add to those which occur when any of the various players has a problem of any type. Any increase in the time needed for a mission tends to increase its cost.

There are instances in which one of the main goals of a space mission is to improve international co-operation, in itself an important political issue. But, when this causes an increase of costs, it is not correct to ascribe the funding of activities aimed at achieving political goals to the space budget.

This not withstanding, co-operation is now an even greater need. However, it must be remembered that, without the powerful incentives of competition, other incentives must be found. Overarching political agreements must be reached to foster the development of clearly stated objectives, around which a strong political will can be obtained, to create a virtuous circle.

Insufficient technology

Were technological inadequacies a contributory cause of the crisis in space activities? At first glance this seems to be ruled out; the technology of the 1960s was able to take men to the Moon and the many design projects for expeditions to Mars are based on today's technology. With current, off-the-shelf technology humankind has sufficient means at least for travelling to our nearest neighbourhood in the solar system. But with this technology we cannot immediately conquer the Moon and Mars, and colonise them.

An example from history can, perhaps, be useful here. There is no doubt that the Vikings in the tenth century had the technology to reach American shores, and Leiv Eriksson actually succeeded. Such an enterprise was repeated and a few colonists from Scandinavia settled for a short time on the American coast. But those settlements were short lived, to the extent that for centuries their very existence was forgotten. Only at the end of the fifteenth century were the Americas 'discovered' and, at that point, colonisation was strikingly fast. In the next few decades new colonies flourished and attracted a large number of people. Thirty years after Christopher Columbus' landing, the people living in the American colonies were numbered in thousands. Thirty years after the first human landing on the Moon, however, no one is permanently living in space.

The ethnologist Ben Finney[11] notes that navigation techniques and the rigging of ships were sufficient to allow people to cross the oceans safely and repeatedly only after the middle of the fifteenth century. Although the right technology is undeniably necessary, favourable political and economic conditions are also required.

If we want to prevent our journeys to the Moon and perhaps later to Mars having the same small impact on history that the American adventure of the Vikings had, we need to develop new technologies so that we can travel safely through space at costs which are compatible with the current economic constraints. When this is so, funding from government sources will no longer be the basic requirement; raising funds via the stock market will be the *modus vivendi*. Whilst the space agencies will deal with scientific research and support major technological advancements, the operational aspects of space activities will be directly in the hands of private organisations. Perhaps something analogous to the way in which the British developed their colonies will be established – a sort of East India Company for the new settlements in the solar system.

We desperately need to improve our space technologies. In the 1980s and early 1990s there was a stagnation – or even a retreat. In the 1970s the Americans had a launcher of the 100 tonnes class, research in nuclear propulsion was well under way, to the point that a prototype nuclear engine had been ground tested, and nuclear reactors were launched into space. In the latter part of the 1990s research was

[11] B. Finney, The Prince and the Eunuch, in *Interstellar Migration and the Human Experience*, University of California Press, Berkeley, California, 1985.

revived, for instance under NASA's Advanced Transportation Plan, and new technologies were tested both on the ground and in space.

The development of new, lighter, stronger and stiffer materials promises to change radically all sectors of technology, particularly space technology. The ensuing increase in performance of space hardware will make practical missions which today are very costly or outright impossible. Materials created using nanotechnologies, by assembling the single atoms, have the potential to bring about a revolution comparable to the introduction of steel in the Industrial Revolution. Single-stage, reusable spaceplanes to reach Earth orbit, and other unconventional launchers designed for routine access to space as described in Chapter 2, used to be considered 'science fiction', but are now on the drawing board.

But we cannot afford to stop our space activities, waiting for some new technological breakthrough to enable us to travel through the solar system with the same ease as we cross the oceans in aircraft today. Because technology usually proceeds in small steps, the innovations which will reduce the costs and times of travel on the routes through the near solar system have already been identified, and are actively being pursued. We need to go forward steadily, making the best use of the technology which we have now and will develop in the future.

COMMERCIAL ACTIVITIES

The commercial exploitation of space is just beginning. In some fields it is already thriving and even becoming profitable. In 1996, worldwide commercial revenues in space reached some US $77 billion and, for the first time, surpassed governments' spending on space activities. In 2001, the figure was about US $83 billion.

Among the commercial space activities are

- telecommunications satellites,
- navigation satellites,
- meteorological satellites,
- Earth resources satellites.

All these tasks are performed by artificial satellites of our planet, operating in the most suitable orbits. If space is compared with an ocean, all present commercial space activities are like fishing near the shore. No commercial activity is performed today in deep space. The exploitation of celestial bodies, such as the Moon or the asteroids, which are potentially important sources of raw materials of great economic value, still belongs to the distant future.

Telecommunications are at present the economic driver of a market for launch vehicles, satellites and related services. These constitute something of an economic flywheel allowing the space industry to overcome the crisis due to the fluctuations of government funding for space activities.

Telecommunications satellites

Telecommunications satellites are the products of the space age which have made the greatest impact not only on our global economic system but also on the way of living and of thinking of many members of the human species. When we speak of the 'global village' we do so because of the extreme ease of communications world-wide.[12] Using radio signals via space we can communicate between two points of the Earth's surface, or even in the air, within a fraction of a second.

Whilst trans-oceanic cables and radio communications have allowed us to transfer information almost instantaneously from one continent to another for much of the twentieth century, the total quantity of information which could be dealt with was severely limited, and many places were difficult to reach. For the majority of people happenings in distant places were regarded as being of little importance; the cost of making a telephone call to someone living on another continent discouraged many from ever attempting to do so. The large number of communications channels via satellites, and so the simultaneous transmission of hundreds of television programmes and of millions of telephone calls, drastically changed this situation

---

[12] See G. Haskell and M. Rycroft (editors), *Space and the Global Village*, Kluwer, Dordrecht, 1998.

and these attitudes. Events which are truly important (e.g., news of a distant catastrophe, such as the Mount Pinatubo volcano in the Philippines erupting in June 1991) or which attract an enormous audience (e.g., the World Cup football final) are broadcast world-wide. They are not really felt to be distant even if they take place on another continent, and the cost of an intercontinental call is nowadays not much higher than that of an intercity one. More and more the Internet is being used to transmit information between individuals, on a global scale.

Telecommunications satellites are usually put into a *geostationary* or *geosynchronous* equatorial orbit, i.e. in an orbit with a period of revolution equal to the period of rotation of Earth, one day. They appear as if they are standing still in the sky, for easy pointing of the transmitting and receiving antennae on the ground. This orbit, whose use for communications satellites was first proposed by Arthur C. Clarke in 1945, is far higher than those of the majority of the other satellites, having a radius measured from the centre of the Earth of almost 42,350 km, corresponding to an altitude of nearly 36,000 km above the surface. There are now 200 satellites in such an orbit. The geosynchronous orbit is becoming overcrowded, to the point that an international convention states that all new satellites using it must carry enough propellant to move them into higher orbits when they are decommissioned at the end of their useful lives.

The availability of communications channels is unprecedented; each satellite in geostationary orbit supplies many TV channels and some tens of thousands of telephone channels. For example, each of the INTELSAT 6 satellites (Figure 1.13), launched in 1990 and 1991 and positioned over the Atlantic Ocean to link Europe and Africa with the Americas, has a diameter of 3.6 m, is 11.7 m long, has a mass of 2,500 kg and can relay simultaneously 24,000 telephone calls and three television channels.

The size of the business required to manage such a communications network is huge. It includes not only the construction and launch of new satellites, but also the operation of the ground stations. INTELSAT 7 satellites are designed for a life of 11 years; they have

Figure 1.13. One of the INTELSAT 6 telecommunications satellites, orbiting above the Atlantic Ocean (J.K. Davies, *Space Exploration*, Chambers, Edinburgh, 1992).

enough fuel on board for 'station keeping', i.e. controlling the satellite's attitude or orientation, for 15 years, in case their actual life exceeds their design life.

Constellations of small satellites for mobile telephones, such as *Globalstar*, should be launched as this book goes to press, but some of the companies involved (e.g. Iridium) have experienced severe financial difficulties. Such constellations have several tens of satellites in orbits far lower than the geosynchronous orbit, at an altitude of about 1000 km or more above the surface. As they orbit, they rise and set at any point on the planet; consequently the communications channel must be switched from one satellite to another, just as happens when the user of a cellular telephone moves from one cell to another.

Navigation satellites

Navigation satellites form one of the best examples of 'dual technology', the civilian use of a technology developed for military purposes, in that case the guidance of ballistic and cruise missiles. To home a missile onto the target with great precision, the guidance system must know its own position with a precision of about one metre. The only way to have such a device capable of operating deep in enemy territory is to use space. Here, the predictability of a satellite's orbit is exploited, or better still that of the orbits of a constellation of satellites, positioned in such a way that at any moment at least three – preferably four – of them can transmit radio signals directly to the flying missile.

The principle is simple. Timed accurately by atomic clocks aboard the satellites, pulses of radio signals are broadcast by the satellites to reach the object whose position must be computed accurately. Having received these at the missile, the distances from the satellites to the missile are computed and, through geometrical calculations, its position can be derived. The NAVSTAR (NAVigation Satellite Time And Ranging) GPS (Global Positioning by Satellites) system was built by the Department of Defense of the United States. Its development started in 1973 to replace an earlier system, the US Navy TRANSIT satellites, operational since 1964. The GPS system consists of a constellation of 24 satellites in six different orbits with a radius of 26,560 km and an inclination of 55° to the Earth's equatorial plane (Figure 1.14). Actually it is a twin system, as each of the satellites broadcasts two signals, one coded, for military use, and the other uncoded, for civilian use. The first gives a precision of a few metres, while

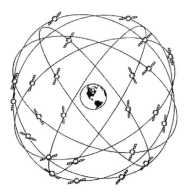

Figure 1.14. Sketch of the orbits of the GPS satellite constellation.

the second is limited in precision to some tens of metres, with a maximum error of 100 m under some conditions.

The Russians developed a similar system (GLONASS, GLObal NAvigation Satellite System). Started in 1982 and completed in 1996, it consists of a constellation of 24 satellites in three different orbits with a radius of 19,100 km and an inclination of 64.8°.

The importance for sea and air navigation of a device which leads to the position of a ship or aircraft being known accurately is obvious. But the fact that the cost of the receiver has reduced to only a few hundred dollars and will come down further is even more important. Today, the owners of many small pleasure boats have a GPS receiver on board; even if in some cases they buy it just as a fashionable gadget, the result is increased safety at sea.

Devices receiving GPS signals are being installed in cars. They show the position of the vehicle on a map projected onto a screen – the driver immediately identifies the best route to follow to reach his or her destination. If the cost of such devices falls to become compatible with other automotive applications, their future as a growing market will be guaranteed. Similar receivers are already widespread for security and rescue systems for automotive use. Devices which continuously broadcast their position computed using GPS signals are especially useful when a car breaks down in a remote area, when there is an accident or when the car is stolen. Their cost is already competitive with that of conventional devices, which cannot match their performance.

To date GPS has not been fully qualified for air navigation but, when it is, it could complement ground-based air traffic control systems and also be used for landing aircraft safely during conditions of poor visibility. It is anticipated that GPS will have a large impact both on air safety and improved profitability through the avoidance of financial losses when flights are cancelled.

Navigation satellite networks were paid for and are maintained by the military, both American and Russian. In the future, civilian systems may be developed, particularly in Europe, where the proposed system is known as Galileo, but it is likely that they will be operated by public organisations.

Meteorological satellites

Meteorology is one of the sciences which has benefited greatly from the ability to observe the Earth from space. Meteorological satellites keep an eye on weather conditions all over the planet, without 'blind spots' as for weather stations on the ground.

Meteorological satellites operate either from relatively low polar orbits, at an altitude of about 1,000 km, viewing in subsequent orbits the entire surface of Earth and taking very detailed images, or from the geosynchronous orbit (Figure 1.15) as used by telecommunications satellites. Then, the satellite always observes the same longitude zone continuously yet over a wide area, but with lower spatial resolution.

Because weather is a global phenomenon and the conditions at any point in the atmosphere are influenced by what is going on at all other points, global observations are of paramount importance, particularly when making forecasts for more than a few hours. Before satellites caused a revolution in meteorology, good forecasts could not be made for more than 24 hours ahead. Now accurate forecasts for four or even five days are possible, thanks also to the rapidly escalating power of supercomputers. Soon it will be possible to predict, reliably, the weather conditions around the world more than a week, or even ten days, in advance.

The World Meteorological Organization, part of the United Nations Organization, co-ordinates the management of data from

Figure 1.15. An image of the Earth's weather systems taken from geostationary orbit (copyright © 2000 EUMETSAT).

satellites on a global scale. Satellites form an even more important part of the integrated network world-wide whereby nations gain access to all the information needed to perform reliable weather forecasts for their region.

The economic value of this network is difficult to quantify, but undoubtedly it is very large. Often the return is in the form of losses avoided rather than in the form of direct gains, distributed across a large number of people. For this reason, while telecommunications satellites are operated mainly by private organisations, meteorological satellites are run by public bodies, such as, for Europe, EUMETSAT. Today, the safety of air and sea navigation, the tourist industry, agriculture and civil defence depend largely on weather information

Figure 1.16. Deforestation in the Rondonia area of far western Brazil, in the Amazon basin, taken from the Space Shuttle during the Space Transportation System (STS) 46 flight in August 1992 (NASA photo).

coming from satellites and interpreted at specialised centres using very powerful digital computers.

Earth resources satellites

These satellites are often large, operating in relatively low orbits around the Earth, between 500 and 1,000 km altitude. The fields of application are wide, ranging from agriculture and cartography[13] to forestry (Figure 1.16), geology, hydrology, soils and volcanoes. Satellites using synthetic aperture radar have a large antenna, and require

[13] World SAT International, *The Cartographic Satellite Atlas of the World*, AND Cartographic Publishers, Finchampstead, 1998.

Figure 1.17. Image obtained by synthetic aperture radar (SAR) of the Isla Isabela in the Western Galapagos from the Space Shuttle *Endeavour* during STS 59 flight, on April 15, 1994 (NASA image).

considerable amounts of electrical power. They have many uses, particularly in the field of oceanography (Figure 1.17).

Studying crops and other vegetation growing on the Earth's surface is the thrust of several programmes, including the American *Landsat* series of satellites and the French SPOT (Système Pour l'Observation de la Terre) series. Each satellite has several instruments aboard tuned to appropriate regions of the spectrum, i.e. observing different colours. For example, the green colour of healthy crops indicates the presence of chlorophyll. Also, regions with a tendency to flood are well monitored from space.

The results of these researches are expected to have large economic returns and to appeal to many types of customers all over the world. Even if to date these satellites have been run by space agencies, the demand for data obtained from space, together with derived, value-added, information, may be so large in the future that Earth resources satellites will be launched and run by private companies.

SCIENTIFIC ACTIVITIES

Scientific research is one of the motives which drove human beings into space; many space missions are still partially or totally devoted to reaching scientific goals. The exploration of space using humans is partly a scientific enterprise, and the research programme of the *International Space Station* bears witness to this. Whilst most scientific missions operate near planet Earth, some have ventured into deep space, to the outer reaches of the solar system (Figure 1.18).

To have a framework for describing the scientific activities which are at present performed in space, the following classification can be useful:

- small scientific satellites,
- fundamental physics research,
- astronomical missions,
- space probes,
- engineering and materials science research,
- observing our Earth.

Small scientific satellites

Space is a large laboratory where scientists can perform experiments which would be impossible on Earth or which, when carried out in space, have special aspects. The main difference between a laboratory on Earth and one in space is that in the latter case experiments are performed in an orbiting vehicle, in the space environment; one special feature is weightlessness or, better expressed, microgravity conditions.

The condition of microgravity inside an orbiting satellite is not experienced because the satellite is far from the Earth – the value of gravitational acceleration at the 300 km altitude of low Earth orbit is only about 10 per cent smaller than that on the Earth's surface.[14] Microgravity occurs because the centripetal acceleration required for

[14] See Appendix B.

Figure 1.18. A montage of images of the planets of the solar system taken by the probes *Magellan, Mariner 10, Viking 1, Viking 2, Voyager 1* and *Voyager 2* (NASA image).

the orbital motion is provided exactly by the Earth's gravitational acceleration. The resulting acceleration is not exactly zero, however, as there are tiny accelerations due to many other causes which perturb the motion of the satellite slightly. These may include the very weak aerodynamic drag of the tenuous upper atmosphere on the satellite, the closing of an electrovalve, the movement of an astronaut, or even his or her sneezing.

Moreover, the acceleration is zero at the centre of mass of the satellite. For all points below the centre of mass (i.e. nearer the Earth) the speed is slightly lower than the correct orbital velocity, and a slight downward acceleration is felt. The opposite applies for points above the centre of mass. Similar tidal effects are also felt by points which are on the right or the left of the orbital plane. These small effects can be exploited to stabilise the orientation of a satellite (stabilisation by gravity gradient).

It is for such reasons that it is more accurate to speak of microgravity conditions within the orbiting vehicle than of weightlessness. Even if zero gravity conditions cannot be reached exactly, they are more readily achieved in a small satellite than in a large one, for example, in the cargo bay of the Space Shuttle or aboard the *International Space Station*.

Some examples of the scientific areas investigated by other satellites are the connection between phenomena on the Sun and near the Earth, the magnetosphere and its Van Allen radiation belts (Figure 1.19), the aurora borealis (northern lights), the properties of the upper atmosphere, tiny particles (the naturally occurring micrometeoroids or the anthropogenically created space debris), and biology.

Fundamental physics research

The microgravity conditions inside an orbiting satellite or spacecraft and the nearly perfect vacuum outside it enable scientists to perform experiments to confirm some of the predictions of those physical theories, such as general relativity, which are considered to be the best available models to interpret the Universe. Or they might refute these predictions, so that we may find some small discrepancies which could start a revolution in our understanding of fundamental physics. Great scientific revolutions often start with very small inconsistencies between some carefully made experimental observations

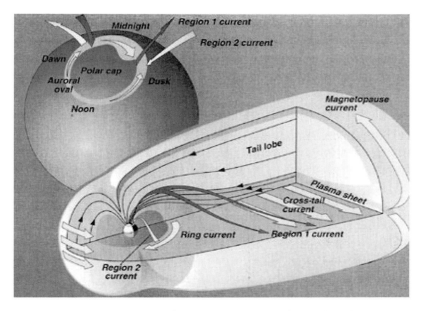

Figure 1.19. Diagram of the Earth's magnetosphere – the volume occupied by the geomagnetic field in space – and, in the upper left, the northern polar region.[15] Large electric currents (millions of amperes) flow through space.

and some predictions of our current theory. The incompatibility between general relativity and quantum mechanics, or the concept of 'dark' matter in the Universe, are still open issues.[16]

Among the fundamental physics experiments in space which have been proposed are those aimed at detecting gravitational waves, checking the equivalence principle between gravitational and inertial mass, and obtaining a deeper understanding of the relationship between matter and space-time curvature. Experiments of different

[15] D.N. Baker, Effects of the Sun on the Earth's environment, *Journal of Atmospheric and Solar–Terrestrial Physics*, Vol. 62, pages 1669–1681, 2000. See also K.R. Lang, *Sun, Earth and Sky*, Springer, Berlin, 1997; S.T. Suess and B.T. Tsurutani (editors), *From the Sun: Auroras, Magnetic Storms, Solar Flares, Cosmic Rays*, American Geophysical Union, Washington, DC, 1998; P. Song, H.J. Singer and G.L. Siscoe, *Space Weather*, American Geophysical Union, Washington, DC, 2001; J.W. Freeman, *Storms in Space*, Cambridge University Press, Cambridge, 2001.

[16] M. Rees, *Just Six Numbers: The Deep Forces that Shape the Universe*, Phoenix, London, 2000; S. Hawking, *The Universe in a Nutshell*, Bantam Press, London, 2001.

types on these issues can also be performed on the ground; a comparison between the results obtained in two such different environments may be very significant.

The precision with which these experiments must be performed in order to obtain significant results is often at the limits of technological feasibility; in many cases very difficult new experimental techniques and new instruments must be developed. In some instances, however, the difficulties involved in space experiments are less than those associated with ground-based experiments, and this makes space an interesting choice for the laboratory.

Astronomical missions

A telescope outside the Earth's absorbing and twinkling atmosphere makes observations which would be altogether impossible from the ground. The atmosphere is transparent only to radio waves and visible light, while it stops most infrared radiation and, happily for us, ultraviolet radiation, X-rays and gamma radiation. Atmospheric turbulence also limits the possibility of precise observations in the visible wavelengths. All branches of astronomy have benefited greatly from space observations; astronomical instruments of all types have been operated in space – large optical telescopes, astrometric satellites such as *Hipparcos*, satellites for infrared, X-ray and gamma ray astronomy, and radioastronomy satellites.[17]

The *Hubble Space Telescope* (HST, Figure 1.20) is without doubt the most famous astronomical instrument in space: its very high cost has already been recouped by results which are of exceptional value in all sectors of optical astronomy. In particular, we have observed nearer the edge of the known Universe than previously, a fact of great importance for cosmological studies. The spectacular images of distant objects[18] (Figure 1.21, among others) taken by the *Hubble Space Telescope* have made a deep impression on public opinion, and have

[17] N. Henbest and M. Marten, *The New Astronomy*, second edition, Cambridge University Press, Cambridge, 1996.
[18] C.C. Petersen and J.C. Brandt, *Hubble Vision: Astronomy with the Hubble Space Telescope*, Cambridge University Press, Cambridge, 1995, M.S. Longair, *Our Evolving Universe*, Cambridge University Press, Cambridge, 1996.

Figure 1.20. Deployment of the *Hubble Space Telescope* from the Space Shuttle *Discovery* on April 24, 1990 (NASA photo).

motivated people to study astronomy. Despite the defects in its initial construction, the installation of 'spectacles' (corrective lenses) demonstrated the ability of humans to work successfully in space, turning a near failure into an outstanding success. Astronauts also replace broken components, perform maintenance tasks and upgrade instruments in space.

By contrast, with simpler instruments, such as on the *Hipparcos* satellite, astronomers estimate with unprecedented precision the distances to thousands of stars. They can then plot these for the first time in an accurate three-dimensional map of that part of our galaxy, the Milky Way, which is closest to the Sun and to the Earth.

The recent discovery of planets orbiting other stars, the so-called extrasolar planets, has proved the existence of other planetary systems. It gives new impetus to the design of astronomical missions to search for planets around some of our nearest neighbouring stars.

Figure 1.21. Image of the M16 nebula, a region of star formation, taken by *Hubble Space Telescope* (NASA photo).

This kind of research is difficult, being at the limits of today's technology. If the human species wishes to find small planets, similar to the Earth, it has to look relatively close to their stars where there might be the right conditions for life to exist. Only with very sophisticated instruments in space might such planets be discovered and, perhaps through spectroscopic analysis, the presence of water, carbon, oxygen or phosphorus (which we consider to be essential for the development of all forms of life as we know it) ascertained.

The Sun[19] is an important object of study for astrophysicists, since it is our closest star. It is an essential energy source for all

[19] K.J.H. Phillips, *Guide to the Sun*, Cambridge University Press, Cambridge, 1992, and K.R. Lang, *The Cambridge Encyclopedia of the Sun*, Cambridge University Press, Cambridge, 2001.

lifeforms on Earth. But it is also the source of charged particle ra-
diations which, from time to time, disturb our telecommunications
systems on planet Earth. Its study has many practical benefits, not
the least of which involves the safety of astronauts and cosmonauts.
Solar activity is the cause of the very variable space weather present
throughout the solar system.

Direct studies of solar activity require space probes. For example, the Sun's polar regions can only be observed by probes which go
out of the ecliptic plane to pass over the poles of the Sun. One example of this type of mission is *Ulysses*, launched by ESA. Other robotic
missions to study the Sun and solar–terrestrial physics were the joint
ESA–NASA *Soho* mission, NASA's *TRACE* mission, and the four
Cluster satellites, one of the cornerstones of ESA's programme, which
were lost in the explosion of the first Ariane V rocket in 1996. A second
attempt at the *Cluster* mission was successful with a pair of Proton
rocket launches during 2000. The Japanese *Yohkoh* satellite has also
obtained remarkable images of the Sun.

Soho is a solar observatory located at the Lagrange point[20] between the Earth and the Sun, i.e. at that point where the gravitational
attractions of those two bodies are exactly balanced. Its aim is the
study of manifestations of solar activity and, in particular, to warn us
of the imminent arrival at Earth of high speed solar wind streams. To
keep a probe at the Lagrange point, which is an intrinsically unstable
equilibrium point, is a technical challenge in itself. In fact, *Soho* rotates around the Lagrange point in a so-called halo orbit. The several
scientific achievements of *Soho* to date demonstrate the success of
doing so.

Space probes

In the last 30 years space probes have reached all the planets of our
solar system, except Pluto, and four of them are now travelling towards interstellar space. The scientific contributions of such probes
to planetology have been outstanding. Very thin discs of dust (like

[20] See Appendix B.

Figure 1.22. Picture of the surface of Venus taken by radar by the *Magellan* probe (NASA image).

Saturn's) have been discovered around Jupiter and Uranus; the surfaces of Venus (Figure 1.22) and Mercury have been studied in detail for the first time, and the composition and pressure of the very dense and hot Venusian atmosphere have been calculated.[21]

Space probes which landed on the Moon and on Mars (Figure 1.23) have provided a great deal of the scientific information which is required to prepare for future exploration by humans. And, to prepare in sufficient detail for Earth people to land on Mars, many probes must still be launched towards the red planet. Nowadays, NASA has a strong programme leading in that direction.

However, neither a number of probes nor a landing for a short visit by astronauts can answer all the scientific questions regarding this intriguing planet. The *Viking* probes which landed on Mars in

[21] All these topics, and many more, are comprehensively discussed in J.S. Lewis, *Physics and Chemistry of the Solar System*, Academic Press, San Diego, California, 1997; and S.R. Taylor, *Destiny or Choice: Our Solar System and its Place into the Cosmos*, Cambridge University Press, Cambridge, 1998.

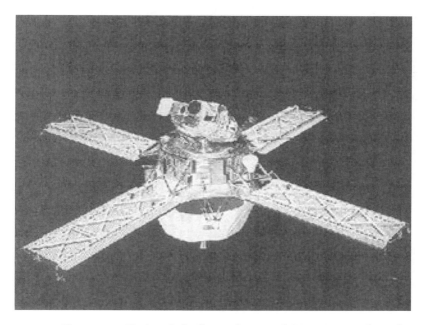

Figure 1.23. *Mariner 9*, the first probe to reach Mars. It entered an orbit around the planet on November 13, 1971, and then sent back data for 349 days. *Mariner 9* showed that the surface of Mars is very irregular, with huge volcanoes and a gigantic valley, which has been named Vallis Marineris (NASA image).

1976 could not confirm the presence or absence of primitive lifeforms, even if the results seemed to indicate that life was completely absent there. But it seems from the images of *Mars Global Surveyor* that in past geological eras liquid water – in the form of rivers and springs – was abundant on the surface of Mars.[22] This, together with the analysis of a meteorite (ALH 84001) coming from Mars which landed in the Antarctic, might indicate that life could also have started there, some billions of years ago, and then disappeared.

An important example of comparative planetology is the study of the greenhouse effect. Venus, the Earth and Mars are three planets

---

[22] M.C. Malin and K.S. Edgett, Evidence for recent groundwater seepage and surface runoff on Mars, *Science*, Vol. 288, pages 2330–2335, 2000.

which are very different from each other, but they are all located in that part of the solar system where, following modern ideas and findings, life can develop and flourish. But Venus is extremely hot and has a very dense atmosphere, while Mars is very cold, with so thin an atmosphere as to be more similar to a vacuum than to what we understand as an atmosphere. It now seems certain that on Venus a 'runaway' greenhouse effect led to such an increase of temperature that almost all the carbon dioxide originally contained in the ground in the form of carbonates was liberated, with a subsequent increase of atmospheric pressure. The opposite seems to have occurred on Mars; the cooling of the planet led to a decrease of atmospheric pressure, by 'fixing' the carbon dioxide in the ground. The decrease of the quantity of carbon dioxide in the air reduced the greenhouse effect, with a further lowering of the temperature, starting a feedback effect which has transformed the red planet into the frozen and barren desert of today. More detailed knowledge of these mechanisms is vital to understand the delicate equilibrium which maintains the conditions on planet Earth which are so favourable to life.

Advances in trajectory design and astrodynamics are an indirect way in which space probes have contributed to the progress of science and space technology. Knowledge of the gravitational fields in the solar system and the technology of trajectory control have now progressed to such a point that a space probe can not only gain energy but it can also proceed in the desired direction by exploiting the gravitational fields of the various planets of the solar system – in that sense it 'navigates' in space, with extreme precision – a precision similar to that of a snooker or pool champion. This is done by getting close to a celestial body different from that of its destination (i.e. by performing a flyby) and using its gravitational pull to extract a relatively tiny amount of the planet's kinetic energy to change the probe's trajectory by a large amount. The *Voyager* probes have visited all the planets outside the orbit of Mars but Pluto, simply by using the gravitational fields of the planets themselves to bend the spacecraft trajectory. Another type of important manoeuvre carried out is where the attractions of two celestial bodies are almost balanced; here, the thrust of a small rocket can cause large trajectory changes. Very complicated trajectories using the notion of 'fuzzy boundaries'[23] have already been found and used. The 'hot topic' of so-called nonlinear astrodynamics was developed to meet the growing exigencies of space probes, and in particular the need for reducing their launch costs. Because the quantity of fuel needed for an interplanetary trajectory is substantially reduced, a smaller launch vehicle can put a spacecraft into its starting orbit around the Earth. This has, however, a price tag attached: trajectories using flybys or fuzzy boundaries take much longer, being slower than more direct, and more expensive, trajectories. The duration of the mission is thus increased, sometimes more than doubled. That increases the costs, having the ground crew operational for longer times, and reliability problems may show up.

Engineering and materials science research

Space exploration has advanced engineering sciences both directly with missions devoted to technological research and indirectly, because new technologies had to be developed to conduct missions which only few years ago seemed to belong to science fiction. Many design and analysis methods, and many new materials, had their origins in the space industry or were developed with funds provided by space agencies. From this viewpoint, in the 1960s and the 1970s the space industry played the same role as the railways did in the nineteenth century. Perhaps the most important contribution of these space missions to engineering science was the perfection of the 'systems approach' to complex problems.

The goal of many missions was the demonstration of new technologies to be used in subsequent missions. But in the last 30 years space technology has become more and more conservative. This is due partly to the shrinking of the available resources, which suggests that the costs and the risks linked to new technologies should be avoided, and conservative, traditional approaches favoured. Another reason is the cancellation of ambitious programmes, which by their very nature required an investment in new technologies, and their replacement by

[23] See Appendix B.

rather routine projects. Another cause can be traced to the generally bureaucratic attitude of the space agencies and particularly their approach to safety. Today, NASA is aware of these problems and many attempts to change its procedures are under way, but it is likely that only a revival of more ambitious programmes and increased funding can bring space technology again to the forefront of technological progress.

The technological research which is being performed in space deals mainly with materials science and biotechnology. In microgravity conditions structural materials and products for use in the chemical and pharmaceutical industries on Earth can be made. Liquids with very different densities can be mixed to create novel alloys by mixing otherwise immiscible liquid metals. Large single crystals can be grown without any contact with the container, and new high-temperature superconductors might be obtained. Such research work is just an example of completely new technologies. Such research work is needed in order to accumulate the know-how which, in the future, will allow true orbiting factories to be constructed; however, these will develop fully only after a large space station, like the *International Space Station*, has become fully operational.

To apply the results of such work to industrial activities in space, the costs of transportation of the raw materials to orbit and of the products back to the Earth must be dramatically reduced. In this respect the Space Shuttle has failed to fulfil its expectations, both for its operational costs and its frequency of launches.

Microgravity research is also important to understand better some processes carried out on the Earth's surface. An example is combustion which, in the absence of gravity and consequently of convection currents, proceeds in a completely different way, to the point of spontaneous extinction.

Observing our Earth

The generic name of NASA's *Mission to Planet Earth*[24] includes a number of space missions aimed at studying our planet in a new

---

[24] The name has now changed to the *Enterprise for Earth System Science.*

way, with an approach similar to that of studying a planet reached for the first time by probes carrying out a scientific mission. In the light of comparative planetology, many of the characteristics of our own planet take on a different aspect and relevance. All of the Earth sciences will benefit from this approach and it should be possible to find solutions to problems which now seem to be insuperable.

The goals stated by NASA for these missions are measurements of the amount of ozone and of other trace chemical species in the stratosphere, and the quantitative study of the polar ice caps, of oceanic currents, of the temperature and the level of the oceans, of tropical rainfall, of land use and of the Earth's thermal balance, involving the energy reflected by clouds and determined by certain substances (e.g., carbon dioxide) in the atmosphere.[25] Longer-term studies will lead to the development of three-dimensional chemical–dynamical models of the atmosphere and its climatology, the characterisation of the biosphere's response to weather changes, the measurement of vegetation globally, deforestation rates and the global fresh-water cycle. Another field in which satellites are changing traditional research methods is the study of the migrations of birds and whales.

In the long run continuous monitoring of our atmosphere, the polar regions and the oceans will be performed with the aim, among others, of developing computer models to predict weather up to two weeks in advance and climate changes on a 10-year scale. Another objective is that of observing, understanding and eventually controlling the use of fresh-water and soil resources at a regional level.

These missions are not very different from those performed by current Earth resources satellites, but the stress is laid more on the scientific aspects of these issues rather than on immediate applications. Studies of the Antarctic 'ozone hole' (Figure 1.24) and of the greenhouse effect are two of the better-known examples.[26] They

[25] See J.E. Harries, *Earthwatch: The Climate from Space*, Ellis Horwood, Chicester, 1990.

[26] J. Houghton *et al.* (editors), *Climate Change 2001: The Scientific Basis*, Cambridge University Press, Cambridge, 2001; National Research Council, *Global Environmental Change: Research Pathways for the Next Decade*, National Academic Press, Washington, DC, 1999. These issues form the research focus of ESA's giant *Envisat* satellite, launched in February 2002.

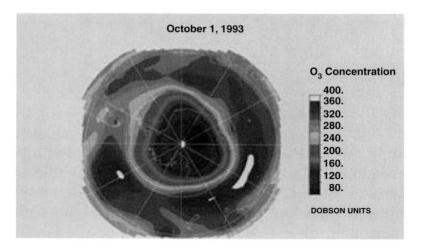

Figure 1.24. The Antarctic 'ozone hole' in a satellite picture of
October 1, 1993. Two hundred Dobson Units correspond to an amount
of stratospheric ozone which, if it were brought to sea level, would
make a layer just 2 millimetres thick (NASA image).

demonstrate that scientific knowledge can produce political actions
aimed at the solution of problems. Laws restricting the production
and the use of chlorofluorocarbons (CFCs) and the growing public consciousness of ecological concerns are early consequences of research
into the Earth's atmosphere.

MILITARY APPLICATIONS

When new technologies are developed, humans soon find some military applications for them. Alternatively, military needs can directly stimulate research which in turn may produce technological
advances. Space technology is no exception and its derivation, at least
in the early stages, from military programmes is clear. Even if the
rockets which launched the early satellites were modified ballistic
missiles, the differentiation between the civil and military sectors
has grown as the space vehicles of today have evolved to have less in
common with military rockets.

Military applications in space can be classified as observation
satellites or armed spacecraft, the latter being more of a controversial
proposal than a fact.

Observation satellites and unarmed military spacecraft
Observation (spy) satellites have been launched since the inception of the space age and their related technologies have shown continuous refinement. Even if the detailed performance of such devices is still secret, it is clear that the spatial resolution of the images is of the order of 10 centimetres. Many different types of observation satellites are being used, some of them very specialised, such as those to detect intercontinental ballistic missiles in their first moments of flight, or to detect nuclear explosions in order to verify compliance with the non-proliferation treaties. These satellites are considered as essential to national security both by superpowers, which operate them directly through their military agencies, and by other countries, which use the information obtained by one of the superpowers through bilateral or multilateral agreements.

In case of conflict, observation satellites can locate and quantify the enemy forces exactly, performing at least in part those tasks traditionally assigned to reconnaissance and intelligence. The 1991 Gulf War showed the paramount importance of space assets in the battlefield (Figure 1.25); they can actually be a winning card in modern warfare.

Observations from space are even more useful in peacetime, above all when this delicate state of affairs is threatened by international tensions. Knowledge of the moves of a potential enemy, and the ability to distinguish between actual hostile intentions and unintentional moves which may be interpreted as hostile, are essential to avoid the dangers of either accidental – and then often escalating – conflicts or an underestimation of the danger.

In the peace keeping and peace enforcing missions in which troops of different nations are involved, often being co-ordinated by international organisations, a detailed knowledge of the situation in 'real time' is essential. Observation satellites are now an important tool among several for preserving world peace.

Observation satellites (e.g., the French *Helios* satellite) are also important in civil defence, particularly in the case of natural disasters, such as large floods, volcanic eruptions or earthquakes. Co-operation

Figure 1.25. Kuwait oil fields (29.5° N, 48.0° E) set alight by the
retreating Iraqi Army during the 1991 Persian Gulf War (NASA image).

between the military agencies of the nations which run the observation networks and the units working on the ground increases the speed of reaction to any emergency situation which is so essential in all civil defence matters. The operations which have to be performed after such a disaster, like damage assessment and reconstruction planning, are made easier by the availability of data supplied by observation satellites.

Armed spacecraft

Armed spacecraft are seen mainly as defensive weapons whose aim is the destruction of enemy observation satellites or ballistic missiles. It is less realistic to think about offensive spacecraft, since ballistic missiles are less readily identified and are more efficient in this role than satellites. Moreover, international treaties forbid nuclear weapons in orbit. An exception is that of Fractional Orbit Bombardment Systems (FOBS), armed satellites which re-enter the atmosphere homing in on their target before completing their first orbit around the Earth. They reach their target from the 'wrong' direction, hoping to slip through potential anti-missile defences. The Soviet Union experimented with these systems with 15 launches, maintaining that the ban on orbital nuclear weapons does not apply to such devices, as they are not strictly operating in Earth orbit.

Even though they are defensive weapons, systems proposed to destroy ballistic missiles (Anti-Ballistic Missiles, ABMs) have attracted much criticism, both for their high cost and for the possible increased risk of nuclear conflicts. Having a space shield capable of stopping enemy ballistic missiles could actually upset a Cold War equilibrium based on mutual fear, i.e. the Mutually Assured Destruction (MAD) policy. The change could even cause one of the opponents, sure in the knowledge of safety from any retaliation, to start a nuclear war. However, such safety would be only a vain hope, because the probability of stopping all the thousands of nuclear warheads which would be launched as a response to such an attack would be extremely low.

The Strategic Defense Initiative (SDI), the project called *Star Wars* by journalists, launched by the Administration of the United States for the construction of a space shield, had very important political consequences, which lie beyond the aims of this book.

The SDI project was not only about missiles to destroy the opponent's ballistic missiles before they hit. It included unconventional weapons, such as powerful lasers, particle beams and electromagnetic guns, mostly based on 'battle stations' in space (Figure 1.26).[27] The space-based weapons try to hit the incoming ballistic missiles when they are most vulnerable, just after launch in their boost phase. Those which escape this first attack are targeted by lasers and particle beam weapons from other battle stations and by lasers on missiles

[27] J.T. Richelson, *America's Space Sentinels: DSP Satellites and National Security*, University Press of Kansas, Lawrence, 1999.

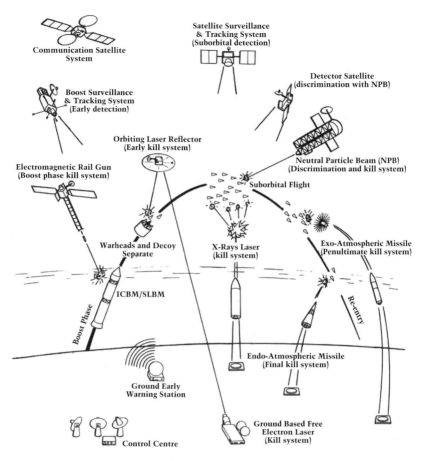

Figure 1.26. Diagram illustrating the concept and elements of a space-based military system both for surveillance and for tracking ballistic missiles, part of the SDIO (R.C. Parkinson, *Citizens of the Sky*, 2100 Ltd., Stotfold, 1987).

and aircraft after the warheads and the decoys have separated, which is a much more difficult task. In the final re-entry phase, the anti-ballistic missiles try to reach those warheads which have found their way through the previous defences, hoping to stop all of them.

This is all very difficult, and the relevant technologies are not yet operational; however, some successful tests of critical components of the system have been carried out, and now the system seems, to some people, to be feasible. The integration of high reliability systems for early warning, battle management and defense against ballistic missiles synthesised in the acronym C4I2 (Command, Control, Communication, Computers, Intelligence, Interoperability) remains the foundation of the space shield concept.

The construction of such anti-ballistic-missiles systems is now limited by international treaties, apart from being made redundant by the end of the Cold War. The ABM treaty signed in 1972 by the United States and the then Soviet Union allowed each of the two superpowers to deploy a single ABM system to defend its capital city. It stated that if either of the two sides intended to abrogate the treaty, a warning of at least six months was required.

The case for a limited space defence system, whose aim is to protect an area from a small attack, carried out with a limited number of weapons, is different. At present the SDI programme has been replaced by the more limited GPALS (Global Protection Against a Limited Strike) programme. This is no longer based on an inter-force structure (the Strategic Defense Initiative Organisation, SDIO), but on separate army, navy and air force structures.

The political debate on the advisability of designing and building a 'space shield' against a limited attack carried out by ballistic missiles is quite heated. The main objection, that of upsetting the global political equilibrium, is no longer viable owing to the limited nature of the GPALS approach. Other objections are its high cost and the difficulty of building a reliable system.

The possibility that an aggressive country (one of the so-called 'rogue states'), or even a terrorist organisation may succeed in obtaining a few ballistic missiles, perhaps with nuclear warheads, is a strong argument in favour of deploying a GPALS system.

At the end of 2001 the Unites States notified its intention to abrogate the ABM treaty. As from July 2002 it will cease to have any effect, and a space shield may be deployed. Russia is looking for a global agreement, which allows it to reduce substantially the costs of maintaining a large number of missiles and warheads. Russia is no longer as strict on the ABM deal. It is possible that broad agreement could be reached among western countries and those of the former Soviet Union to include the space shield in a wider context of treaties on global security. It might even be that a common system could be developed.

Most of the innovative weapons studied in the SDI programme require large quantities of energy to be generated aboard satellites, and knowhow in this field is very valuable for a variety of civilian space applications. One of the main goals of SDI research was the development of high-power lasers; in space these could be used as propulsion and long-range communication devices. A further application of the technologies developed for the *Star Wars* project is defence against large meteorites and small asteroids whose orbits bring them so near to the Earth that a collision – with disastrous consequences for all aboard planet Earth – could occur.[28]

Although the SDI system is generally referred to as a space shield, i.e. a purely defensive option, the technologies which must be developed could also be used offensively. A battle station might aim its weapons at ground targets (perhaps the control centres of the opponent's space shield) or aircraft. Using energy beams, such an attack leaves only milliseconds for a response, upsetting the MAD strategy. When those technologies are ready, there will undoubtedly be a push to develop some form of protection against them – a new shield against a new sword.

A number of experiments in which a satellite destroys another satellite have been performed in the past, mainly by the Russians. Such experiments are very harmful, creating as they do a large number of small, high-speed fragments in space. These add to those produced in an accidental explosion of the remaining fuel in the last stage of a rocket, or to other intentional explosions of decommissioned military satellites, aimed at preventing them from falling into the hands of a potential opponent. Fortunately, most of the fragments eventually fall into the Earth's atmosphere, and are burnt up. But the problems

[28] See Chapter 4.

caused by such 'space debris' are becoming more and more serious, particularly for the most popular, busiest orbits.

Devices to jam enemy radio communications or computer systems can be considered as lying between observation satellites and armed spacecraft. However, there is little written about them in the public domain.

SPACE AND THE DEVELOPING COUNTRIES

One of the most common objections to space activity is that it is insane to waste large amounts of money in space when those funds could be used to solve the problems which still plague the Earth, in particular those affecting a large part of humankind still living in very poor conditions.

One answer to such an objection is that the resources used for space activities are too small to contribute significantly to the alleviation of any one of those problems. Actually, the total cost of humankind's space activities is not so high as it seems at first sight. Compared with military expenditure, space is a low-cost enterprise. Around the world, military expenditure is about US $800 billion per year – for that money more than two thousand probes could be sent to Mars each year! Large scale civil projects, like building motorways or railways, have costs comparable with space costs. With the money which the Italian (or French, or British) people spend to celebrate each New Year, it would be possible to send six space probes to Mars!

Moreover, there is no way of being sure that what is saved by the reduction of space activity would actually be spent to help the poor. Also, the value of the object launched into space is just a small fraction of the overall cost of any space mission. The largest part of the funds is spent on equipment and salaries for technological and scientific research, which have a fall-out in all fields of human activity, thus perhaps helping to solve the problems mentioned above. These answers are, however, rather weak and do not focus on the essence of the matter. While such technological transfer does exist, it is not a very efficient way of investing resources in research to obtain results in completely different fields.

Those who support the space imperative give a tough answer – space exploration is a primary duty of humankind, who must not let themselves always be distracted by problems, even by the very serious ones. The betrayal of this task will have severe consequences for all. The space imperative deals with long-term perspectives. Of course, it could be counter-argued that it is wiser first to solve the problems plaguing our planet and then to begin our expansion into space without leaving behind the poor and underdeveloped members of humankind.

The answer put forward by supporters of the space option is more convincing: the problems of poverty and underdevelopment will never be satisfactorily solved in a purely terrestrial perspective, and not with a levelling of the living conditions of everybody to a rather low standard. The results of the well-known Report to the Club of Rome, *The Limits to Growth*, are very clear: in all simulations the living conditions get worse and the population finally decreases, after a strong initial increase (Figure 1.27). Put bluntly, this means a strong increase in human mortality. There is no doubt that these simulations have been criticised and many of the report's predictions have already proved to be incorrect. But it is impossible to overlook the fact that the simultaneous presence of overpopulation, pollution and insufficient resources will jeopardise our future on Earth.

Moreover, some simulations did show that expansion into space and the exploitation of extraterrestrial resources can make things better, even solving completely the difficulties caused by the limited resources of our planet. However, any delay in implementing the space option could have very severe consequences. The reduction in available resources, which could be felt from the early years of the twenty-first century, would slow down that very expansion into space which would, in the end, make the space option itself useless. Reading the articles of many of the proponents the authors feel some urgency, a last chance not to be missed.

Perhaps the best answer to the objections raised against space activity in the name of the impoverished people of the Earth is that the developing countries are amongst the keenest users of the services

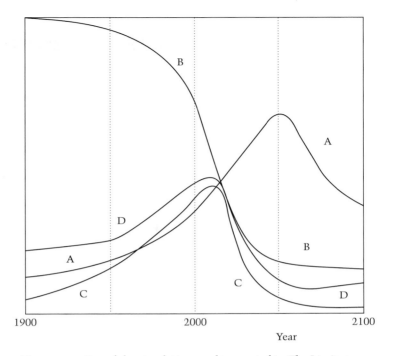

Figure 1.27. One of the simulation results reported in *The Limits to Growth*. Trends of population (curve A), natural resources (B), per capita industrial product (C), and per capita food availability (D) from 1900 to 2100.

obtained through space systems. Further, the United Nations Organisation has a Committee (COPUOS, COmmittee for the Peaceful Uses of Outer Space, which organised the 1999 Unispace III Conference) which encourages the space activities of countries without their own space agencies to be co-ordinated with those of other countries.

The distinction between the developed and developing countries is increased by the cycle linking the lack of infrastructure to poverty and poverty to the impossibility of improving their infrastructure. Such a cycle can be broken only by using space technology. The services available from space systems are important for all countries but are essential for the developing countries,[29] which have no

[29] See G. Haskell and M. Rycroft (editors), *Space and the Global Village*, Kluwer, Dordrecht, 1998.

alternative, owing to their lack of infrastructure. This is true for the exploitation of material resources, the reduction of the consequences of natural disasters, mainly floods, the improvement of the quality of life and a better use of human resources. Telecommunications satellites enable large-scale programmes in the fields of telemedicine and education to be started, from learning the alphabet to vocational training. They provide rapid access to the information highway, even from remote places. China, India and Indonesia are, perhaps, the best examples in this field to date.

Finally, it must be remembered that the cost of the actual space equipment, including the launch cost, is smaller than the cost of the ground infrastructure, which is generally not located in remote places. Whilst the primary needs of developing countries are food, water and electricity, space systems only supply information. Information from space is a powerful tool for obtaining food, water and power only if the appropriate ground infrastructure is in place.

ROBOTS OR HUMANS IN SPACE

The advisability of human beings participating directly in space adventures has stirred up and continues to cause many debates. There is no doubt that the presence of people on board a space vehicle makes its design much more complex and challenging. It produces a large increase in costs. It has been thus ever since the beginning of the space age, and is even more so today.

First, the requirements for safety of the space vehicles are greatly increased. Today, many ask themselves whether, in the case of robotic missions, it would be better to increase the number of space probes which are launched, decreasing their reliability. The cost reduction and the increase of the results/costs ratio which can be achieved in this way can be important. It is inevitable that the very high reliability which is needed when humans are on board increases the costs greatly. Secondly, it also causes large lead times between the initiation and the completion of any mission.

The performance of the life support systems required by the people aboard a space vehicle must be guaranteed. However, not only do they produce minimal comfort but also they are heavy, bulky and costly; their complexity increases for long-duration missions. The miniaturisation of instruments and all electronic devices makes things worse from this viewpoint. As technology advances, robotic probes become lighter, smaller and more convenient than manned vehicles. This leads to a reduction of the size, and cost, of the launch vehicles. Furthermore, advances in electronics and computer science allow increasingly complex tasks to be entrusted to robots.

However, experience has shown that the presence of humans in space is popular with the public. It is also useful; there are many cases when only direct intervention by an astronaut corrects the mal-function of an automatic device. The ability to react to unexpected situations and the ability to perform a wide variety of tasks are two human characteristics which are precious in space missions. Astro-nauts and cosmonauts have proved that they can adapt to conditions of weightlessness and work in space without encountering too many problems. The operations to repair and to upgrade the *Hubble Space Telescope* are perhaps the best example, but they are just two of many (Figure 1.28).

This is even more true in the case of deep space missions. If the human exploration of Mars is a very difficult enterprise, robotic explo-ration is not much simpler. Any automatic probe moving on the sur-face of that planet must work autonomously. While in the case of the Moon it is possible for someone on the ground to 'tele-operate' a probe, as the two-way link time is only a couple of seconds, the same cannot be done to Mars. Several minutes elapse between the instant when the camera of a rover probe detects an obstacle in its path and that when the course correction commands arrive from the Earth.[30] The automatic vehicles which crawl over the surface of the planet, such as the *Sojourner* (Figure 1.29) rover of the *Mars Pathfinder* mission of 1997, are very slow and must have a good operational autonomy.

---

[30] The time needed for a radio signal to travel between Mars and the Earth ranges from 4.4 minutes, when the planets are nearest to each other (Mars 'in opposition' with the Sun), to 21 minutes when they are on opposite sides of the Sun (Mars 'in conjunction' with the Sun).

Figure 1.28. An astronaut performing ExtraVehicular Activity (EVA) to repair the *Hubble Space Telescope* (NASA photo).

However, the *Sojourner* rover was manually tele-operated in spite of the long time needed to receive the control inputs. The behaviour of robots faced with unexpected events is still not well known.

Many of the promises of artificial intelligence are still far from being fulfilled. The construction of machines simulating human logical reasoning moves towards ever more distant dates, as does the

Figure 1.29. The robotic rover *Sojourner* on the surface of Mars (NASA image).

human exploration of Mars. The more that the performance of computers improves, the more it is realised how difficult it is to build machines which display actual logical abilities.

Although nobody has yet succeeded in defining exactly what intelligence is, or perhaps mainly for that reason, today the term 'intelligent' is applied to a variety of situations – intelligent machines, intelligent structures, intelligent weapons, and intelligent suspensions in motor vehicle technology. And perhaps that is also with good reason, as these devices now are capable of performance which was unthinkable a few years ago. But the term 'intelligent' must be properly understood: who has not used it when seeing a dog which, when ordered to give its paw, presents its right front leg? Many machines are without any doubt intelligent in this context: the intelligence expected from a human being is completely different.

Similar considerations apply in the industrial world. Many discussions have taken place on fully automated factories, in which all operations are performed without any human intervention, and forecasts of the complete substitution of workers by robots in many production arenas have been made. Today, these perspectives are being revised. All machines, even the smartest ones, must cooperate with humans and classical robots are often unsuitable for such a task. The word 'cobot', from 'collaborative robot', has been invented to designate an intelligent (in the above sense) machine capable of helping a human operator without replacing him or her.

A similar trend is also apparent in the space field. Tasks which were in the past entrusted only to machines are now performed by human beings, sometimes with the aim of using simpler and less costly devices, sometimes to obtain better performance. Attempts in this direction, even clumsy and dangerous ones, can avoid costly automatic devices. A much publicised accident occurred to the *Mir* space station when a *Progress* automatic cargo vehicle failed a docking attempt under manual guidance, hitting a solar panel and causing much damage. In that case the docking manoeuvre was performed manually, to cut costs, without providing the pilots with the required instrumentation to perform it safely.

In other cases to 'put the person in the control loop' is a welcome simplification, which lowers the cost of a mission without compromising its safety. The lunar probes planned by ESA, for instance, will be 'piloted' by a human operator on the ground to a greater extent than previous lunar probes. In this perspective, the added costs due to the presence of people on board a spacecraft can at least partially be compensated for by a reduction of the cost of the control systems. Many operations, which were meant to be performed under completely automatic control, can be performed more efficiently by astronauts, perhaps helped by their 'cobots'. The human–machine relationship, which sometimes tends to become conflictual, must evolve towards a closer co-operation.

An example of this human–machine co-operation is the Mars Outposts approach to Mars exploration, recently launched by the Planetary Society and discussed in more detail in Chapter 6. Here, a number of robotic research stations, equipped with permanent communications and navigational systems, would be sent to the red planet. They would perform research, and establish the infrastructure needed to prepare future landing sites and return vehicles for the exploration of Mars by humans.

In the most difficult environments, as on Venus or Jupiter where humans could not survive, robots could be controlled by human beings located in the shirt-sleeves environment of an orbiting spaceship or from a base on a Jovian moon.

A reduction of the cost of launching payloads into Earth orbit – of the cost of getting out of the Earth's gravitational well – is essential. This will only result from marked progress in new launch technologies, and will make it easier for human beings to participate directly in space exploration. And there is a cascade effect: the more humans who live in space and settle on other celestial bodies, the fewer people and the less materials to be launched into space from the Earth. What is required in our generation is a 'bootstrap' effort, which will slowly gain momentum.

If space is to be more than a place to build automatic laboratories and to start some industrial enterprises in the immediate vicinity of our planet, the presence of humans is essential. Humans must learn not only how to work but also how to live in space for many years. They must learn how to voyage through space towards destinations which will be not only scientific bases but also places to live. In other words what humankind can do – and in the future could decide to do – is to colonise space, and not only to send robotic devices to explore it. This is the opinion of many, and the question contained in the title of this book focuses mainly on this issue. If space is a frontier,

Figure 1.30. The Earth rising above the lunar surface, as seen by the *Apollo 11* astronauts (NASA photo).

that frontier must see the presence of people in space. This important statement has many psychological implications; among the motivations which pushed humankind to explore and to colonise new lands, irrational ones have always been present. Human curiosity and the quest for adventure have always played important roles. Surely they always should.

The result of exploring and settling in space will be a deep change in human culture and in the views which humankind has not only of the Universe but also of itself. And this process is already under way. Many think that the images of the Earth taken from the Moon (Figure 1.30) are some of the most important results of the

Apollo programme. To see our planet from that distant viewing point gave humankind a new consciousness of the fragility, the smallness, and the unity of our world.[31] These impressions were among several which triggered a general awareness of the need to protect and preserve our planet. For it is the place in the solar system most suitable for us and, above all, it is the only planet we have – at least for now.

[31] C. Sagan, *Pale Blue Dot: A Vision of the Human Future in Space*, Headline Book Publishing, London, 1995.

2   The gateway to space

A reliable rocket is the first requirement for a successful launch into space. It transforms the inherent chemical energy of the fuel into kinetic energy of motion. Higher performance, and environmentally friendly, propulsion devices are needed to launch equipment into space at much lower costs than at present.

THE TWO COSMIC VELOCITIES

The possibility of putting objects into orbit around the Earth was understood by Sir Isaac Newton as a consequence of the law of gravitation, which he discovered. He described a conceptual experiment, as was usual at his time, which can also be useful today to understand what an artificial satellite is.

Imagine that a gun is put on the top of a high mountain and a bullet is shot in the horizontal direction (Figure 2.1, taken from the original illustration by Newton[1]). To use Newton's own words, 'Let AFB represent the surface of the Earth, C its centre, VD, VE, VF the curved lines which a body would describe, if projected in a horizontal direction from the top of a high mountain, successively with greater and greater velocity...'. The first trajectory will be a parabola and the bullet will fall to the ground at point D. If the initial velocity is increased, the bullet will fall at a greater distance, at point E, always following a parabolic trajectory. Actually the trajectory is not exactly parabolic: if the fact that the Earth is spherical is taken into account, the trajectory is an arc of an ellipse. Newton goes on: '... and augmenting the velocity, it goes farther and farther to F and G; if

[1] Isaac Newton, *A Treatise of the System of the World*, London, 1728, and Dawsons, London, 1969.

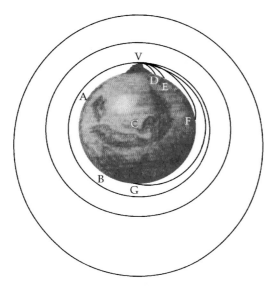

Figure 2.1. Conceptual experiment to explain how a satellite achieves its orbit. Increasing the initial velocity, the bullet shot from point V falls at an increasing distance (D, E, F and G) and then enters a circular orbit around the planet (from the original figure in Isaac Newton, *A Treatise of the System of the World*, 1728). The outermost orbit shown is elliptical, rather than circular.

the velocity was still more and more augmented, it would reach at last quite beyond the circumference of the Earth, and return to the mountain from which it was projected'.

This experiment, as Newton pointed out clearly, is only a theoretical one, as air drag will slow the bullet down making it fall to the surface sooner. It is impossible to find mountains high enough or guns powerful enough to perform the experiment. But it throws light on a fundamental fact: to put something into orbit the key factor is to provide it with a high enough velocity. The bullet remains in orbit because the centripetal force, due to the curvature of the trajectory, is exactly provided by the gravitational attraction of the body. This is also the reason why something or someone within an orbiting satellite experiences conditions of weightlessness.

The higher the orbit, the weaker the gravitational attraction of the Earth and, as a consequence, the lower the speed required for a circular orbit. The orbital velocity at the surface of the Earth, which is sometimes referred to as the *first cosmic velocity*,[2] is about 28,500 km/hr. It decreases slowly with increasing altitude but has essentially the same value for all low Earth orbits (LEO), i.e. orbits with an altitude between 200 and about 1000 km above the Earth's surface.

If the velocity does not exactly match the orbital velocity at the altitude where it is injected into orbit, the resulting trajectory is an elliptical orbit, with the centre of the Earth at one of the two foci of the ellipse. If the speed is too low, the orbit lies within the circular orbit, and may intersect the surface of the Earth (as do the orbits from VD to VG in Figure 2.1). If the speed is appreciably higher, the orbit is outside the circular orbit. The higher the speed, the more the orbit is elongated, and the satellite reaches greater distances from the Earth.

Now imagine that the gun is put in a vertical position to shoot directly upwards. The bullet will slow down during its ascent until it stops and then starts falling. By increasing the initial velocity it will reach a greater altitude before descending. But the gravitational field which the bullet experiences in its upwards flight becomes weaker: there is a value of the initial velocity for which the bullet, although continuing to slow down, will never fall back again. This velocity, such that the point at which the speed is reduced to zero is infinitely far from the surface of the planet, is the escape velocity, sometimes referred to as the *second cosmic velocity*. For Earth its value is about 40,250 km/hr; it is $\sqrt{2}$ times the *first cosmic velocity*.[3]

Note that if a spacecraft is launched with this escape velocity it slows down, reaching a very low speed (theoretically equal to zero). A greater speed is needed to perform any actual mission in deep space. The speed at which the spacecraft travels through space, once it is sufficiently far from the Earth, is the so-called hyperbolic excess speed (see Appendix B). This grows with increasing launch velocity

---

[2] The velocity increase Δv needed to launch an artificial satellite is greater than 28,500 km/hr, as some energy is also required to take the satellite up to the altitude of the orbit.

[3] A simple proof of this is given by D. Shapland and M. Rycroft, *Spacelab, Research in Earth Orbit*, Cambridge University Press, Cambridge, 1984.

from the Earth. This explains an apparent contradiction: at a speed of 40,250 km/hr the Earth–Moon distance (384,000 km) should be covered in nine and a half hours and the minimum Earth–Mars distance (78.7 million kilometres) in about 81 days, while the times actually required for missions of this type are about two and a half days and 230 days, respectively. Units of measurement are explained in Appendix A.

ROCKET PROPULSION

Perhaps without knowing it, Jules Verne followed the experiment of Newton in his novel *From the Earth to the Moon*, using a gun. This is practically impossible, not only because no existing gun is powerful enough, but also because awesome accelerations are needed to reach the escape velocity from Earth within the length of the barrel. As the barrel of the gun imagined by Verne, the Columbiad, was 275 m long, the acceleration time would have been 0.05 s and the acceleration more than 22,000 times the gravitational acceleration on Earth. This means that, if the characters of the novel by Verne had a weight of 70 kg at the moment of the launch, they would have been squeezed to the floor by a force of 1,500 tonnes (about three quarters of the weight of the Space Shuttle at launch). No known substance can survive such accelerations intact.

A propulsion device which accelerates the space vehicle more gradually is needed. It must attain the required speed with an acceleration of about three or four times the gravitational acceleration, if there are to be people on board the vehicle. Up to now, all space missions have been launched by rockets using liquid or solid propellants as the fuel.

Jet propulsion is based on the fact that if a vehicle ejects matter in a given direction, a force acts on it in the opposite direction. Having stated his third law, Newton described another of his conceptual experiments, about a cart pushed by the thrust due to a jet of steam. Two of the foundations of space locomotion were thus laid in the seventeenth century, but more than 200 years had to pass before these ideas could be put into practice. As recently as 1914, a *New York Times* journalist, writing about the attempts of Robert Goddard to build a liquid propellant rocket, wrote that Goddard (after whom the NASA centre in Maryland is named) did not know elementary physics. Beyond the Earth's atmosphere the rocket could not work as there was no air against which the jet could exert its force. The authors of this book have several times heard the same, wrong 'common sense' reasoning. The *New York Times* recognised this error in an article published in the 1960s.

But the lack of air does create a problem: in an aircraft jet engine a fuel, usually kerosene, is burnt using the oxygen from the air breathed in by the machine. Outside the Earth's atmosphere the engine must carry not only the fuel but also the oxidiser, gasoline and liquid oxygen, for instance, or hydrogen and oxygen, both liquids. A rocket is significantly different from an air-breathing jet engine.

As the quantity of propellant needed to produce a given thrust for a given time decreases with increasing velocity of the gases which constitute the jet, the hydrogen–oxygen combination, which produces the largest jet velocity among the commonly used propellants, is the most convenient. Unfortunately, the density of liquid hydrogen is very low and so it takes up much space: this is the reason for the very large external tank of the Space Shuttle (Figure 2.2).

One of the objections which led many to think that it was impossible to reach the speed required to orbit the Earth is based on the fact that the speed which a rocket can reach depends on the quantity of fuel that it carries. A huge quantity of propellant is needed to attain the speed required to enter low Earth orbit, more than 90% of the mass of the entire vehicle at lift-off. This means that less than 10% is left for the structures, the engines, the control devices and the payload.[4] It is a serious problem but one that has a solution – multistage rockets. One rocket accelerates another smaller rocket to a moderately high speed and then, once its propellant has all been burnt, detaches and

---

[4] See Appendix C.

Figure 2.2. Launch of the Shuttle *Endeavour*, with its external tank and solid propellant boosters (flight STS 47, September 12, 1992, carrying the Spacelab, NASA photo).

falls back to the Earth's surface. The second rocket fires and further increases the vehicle's speed. Its payload can be a third rocket, and so on to reach even higher speeds. Multistage rockets are essential for reaching speeds high enough to put objects into Earth orbit or to leave our planet.

Apparently, a rocket engine is a simple machine: a propellant consisting of a combination of a fuel and an oxidiser, solid or liquid, is burnt in a combustion chamber. The products of combustion are expanded in a nozzle to form a jet, whose velocity is as high as possible. But practically it is a highly complex machine, particularly if liquid propellants are used, as many auxiliary systems, such as turbopumps, and control and guidance devices must be present.[5]

The thrust must not be too great, because the acceleration imposed upon the vehicle cannot be larger than some permitted limit. But it must also not be too small: when a rocket lifts off vertically a part of the thrust counterbalances the weight of the machine and is completely useless for acceleration. The thrust must be greater than the weight – the greater it is, the smaller the proportion which does not contribute to the acceleration, important on the total balance sheet of the fuel consumed. From this viewpoint, the acceleration should be as large as possible. Clearly, a trade-off between the two factors is required, but only chemical rockets are at present able to supply high enough values of thrust to leave the surface of our planet. For the first stage, however, it is possible to use air-breathing jet engines; these use the oxygen in the air as oxidiser, and there are some projects for their use as space launch vehicles.

A chemical rocket burning liquid hydrogen and oxygen is the cleanest engine which we can think of – the combustion product is only water vapour. There are none of the dangerous (to the environment) combustion products (e.g., carbon dioxide, soot) of aircraft or car engines, or of the boilers of power stations.

Thermal nuclear rockets (TNR)[6] could substantially reduce the mass of the launcher, and hence the cost of sending a payload into orbit. Their basic principle is to use an on-board nuclear reactor — instead of the combustion of the fuel and oxidiser — to heat a fluid, producing a high speed jet. In spite of the research work performed in the 1960s and 1970s, no nuclear rocket engine has ever been flight

[5] See Figure 1 in Appendix C.   [6]See Appendix C.

tested. Most people, even in the space community, think that the danger of radioactive leakage into the atmosphere is too large to contemplate their use for lift-off from the Earth's surface. However, they believe that nuclear rockets could be used safely in deep space, where radiation is naturally present so that such leakage is of no concern.

However, recent studies by NASA suggested that it is safe to use nuclear propulsion, even for launching a rocket from the Earth. A new reusable launcher with a nuclear second stage, which would be ignited after the boosters have taken the vehicle to about 10 km altitude, is now under preliminary study.

BEYOND THE SPACE SHUTTLE

The STS (Space Transportation System) Space Shuttle[7] is essentially a rocket able to reach low Earth orbit, and then re-enter and land as an aircraft. Actually, it lands like a glider, since it does not use its engines in this last part of the flight. The vehicle lifts off under the thrust of both of its own engines, burning liquid hydrogen and oxygen stored in the external tank, and of two huge solid propellant boosters. Once their fuel is exhausted they are released from the vehicle. The two boosters have parachutes to touch down softly on the sea, and are then recovered. The external tank is the only non-reusable part of the STS.

The total mass of the STS at lift-off (Figure 2.2) is about 2,000 tonnes. It can put slightly more than 24 tonnes into low Earth orbit, i.e. a mass of 1.2% of the total lift-off mass. It can carry a crew of seven, but in an emergency it can accommodate ten people. The cost of an STS launch is currently about US $30,000 per kilogram put into low Earth orbit.

Its performance is exceptionally good from a technical point of view. When its design was started, many, even in the aerospace community, had doubts about the feasibility of such a machine. Whether a spaceplane could withstand the extremely large thermal stresses

---

[7] F.H. Winter, *Rockets into Space*, Harvard University Press, Cambridge, MA, 1990; V. Neal, C.S. Lewis and F.H. Winter, *Spaceflight, A Smithsonian Guide*, Macmillan, New York, 1995; D.M. Harland, *The Space Shuttle: Roles, Missions and Accomplishments*, Wiley-Praxis, Chichester, 1998.

during the re-entry phase was uncertain, as was whether it could operate in space for a long time without significant damage. The fact that three Space Shuttles are still operating reliably after 20 years is an outstanding technical success. The *Challenger* tragedy must not detract from the success record of the Space Shuttle: the very fact that in more than 100 flights there was one catastrophic failure means that its reliability is higher than 99%. The accelerations which the crew experiences, both at liftoff and during re-entry, are lower than those typical of non-reusable rockets. Thus, people who are not trained as astronauts can have access to space.

However, the technical achievements must not let us forget that the Space Shuttle has been something of a failure from the commercial and economic viewpoint. The cost of each launch is much higher than had been hoped for in the design phase, and the time required to prepare for another launch is too long. The Space Shuttle has failed to provide customers with an easy, frequent and relatively cheap launch service. To penetrate the space frontier, new launch vehicles are definitely required.

Since the *Challenger* disaster of 1986 one more vehicle has been built and now four Space Shuttles are operating – *Atlantis, Columbia, Discovery* and *Endeavour.* A sixth shuttle, *Enterprise,* was used for the first unmanned tests in which it was carried on the back of a modified Boeing 747 and then released to glide to a normal landing. Nowadays, it is a technically obsolete machine, and the four vehicles now operating are ageing. It is a common opinion among the *cognoscenti* of propulsion that it should be replaced by a more modern transportation system without losing further valuable time – it will take many years to develop such a machine from scratch.

The only other reusable launch vehicle is the Russian *Buran.* It is very similar to the Space Shuttle, regarding both its operational characteristics and its size and general aspect. Together with the rocket Energia, which acts as a launch vehicle, the total mass at liftoff is 2,400 tonnes, with a payload of about 30 tonnes. The Energia rocket and the *Buran* spacecraft were developed separately, and the former can be configured in different ways. It can take up to 200 tonnes into low Earth orbit, in a configuration which has, however, never been flight tested.

The Energia–*Buran* complex (Figure 2.3) has been used only once, in November 1988, with an orbital unmanned test flight. Although the flight was very successful, it has never been used again. It is an excellent example of a casualty caused by the financial troubles of the Russian Space Agency.

Europe has been very active developing and using the Ariane IV and V launch vehicles[8] in the 1980s and 1990s. The European Space Agency (ESA) has also started a programme named FESTIP (Future European Space Transportation Investigations Programme) to develop new reusable orbital vehicles. The French Space Agency (CNES) made a proposal for a spacecraft which should have lifted off vertically using a booster, with a general layout not dissimilar to the American and Russian shuttles, but of smaller size, named *Hermes*. The project was approved by ESA in 1987, together with the rocket Ariane V to launch it. After some downsizing, the project in 1991 was based on a low-Earth-orbit capability of 1,000 kg and the spacecraft had to carry a crew of three. It was a very interesting machine, of a size such that it complemented rather than competed with the American and Russian shuttles. Unfortunately, the project was then cancelled.

The Japanese Space Agency NASDA has been working on a small automatic spacecraft, of the same class as the *Hermes*, to be launched by the *H2* rocket, which became operational in 1994. The spacecraft, named *Hope* (*H2* Orbiting Plane), would have operational characteristics similar to those of the Space Shuttle or of *Buran*, with the important difference that, at least initially, it is not meant to carry people. The payload in low Earth orbit should thus be larger than that of the *Hermes*, from 3 to 5 tonnes. The latest design is of lower mass, and called *Hope-X*.

Some proposals for two-stage, totally reusable, spacecraft have been made. They are machines of a completely different type,

[8] B. Lacoste, *Europe: Stepping Stones to Space*, Orbic, Bedfordshire, 1990.

Figure 2.3. The *Buran* spacecraft ready for lift-off in 1988, with the Energia rocket (J.K. Davies, *Space Exploration*, Chambers, Edinburgh, 1992).

practically a winged spacecraft launched from an aircraft. The larger vehicle, which acts as a first stage, can reach a speed[9] of Mach 6–7, while the smaller one, launched from the first at altitude, can reach low Earth orbit. The first stage is usually powered by air breathing engines, turbojets for take-off and ramjets (perhaps scramjets, i.e. supersonic combustion ramjets, which will power the X-34 aircraft) for the subsequent acceleration.

A vehicle of this type is the German *Sänger*, a project started in 1988 aimed at developing a launch vehicle able to take off from and land at any major European airport. The vehicle had a take-off mass of 370 tonnes, including a 110 tonnes second stage. Two different second stages were studied: the Horus (Hypersonic ORbital Upper Stage), able to put 3 tonnes of cargo and three people into low Earth orbit, and the Cargus (CARGo Upper Stage) capable of carrying 7 tonnes without a crew. The first stage had to be powered by liquid hydrogen turbo-ramjets while the second stage would be propelled by hydrogen–oxygen rockets. The project has been cancelled.

Particularly interesting projects are those based on the SSTO (Single Stage To Orbit) concept. From what has been said above, it might be thought that to reach satellite velocity at least two stages are needed and so a SSTO launcher should be impossible. Actually, advances in materials and engine technology are making it possible to overcome the great practical difficulties involved, even if some rocket specialists are still sceptical on the matter. Some SSTO projects are under way. Certainly, a SSTO spacecraft would carry into low Earth orbit a smaller payload than an equivalent multistage vehicle, but its simpler layout (for example the smaller number of engines, with all their ancillary equipment) and – above all – its lower operational complexity would initiate a true revolution in the very way in which spacecraft are launched and operated.

The use of air-breathing engines for the first part of the flight would make it far simpler to build a SSTO, as far less oxidiser needs

[9] The Mach number is the ratio between the speed of the aircraft and the speed of sound. At sea level the speed of sound is about 330 m/s (1,190 km/hr); a Mach number of 6 therefore corresponds to about 7,000 km/hr.

to be carried on board. In 1986 the Hotol (HOrizontal Take-Off and
Landing) programme was started in Great Britain, with the goal of
building a SSTO spaceplane with a take-off mass of 250 tonnes,
powered by liquid hydrogen turbojets and hydrogen-oxygen rockets.
In 1990 the project was completely transformed into the British–
Russian AN 225–Interim Hotol project, with a Hotol spacecraft, of re-
duced size, carried by the subsonic Antonov 225 aircraft (Figure 2.4).
The take-off mass of the AN 225 is 600 tonnes, while that of the
Hotol would have been 250 tonnes. The programme was cancelled in
1992.

The United States developed the Space Shuttle II project, which
would have been a spacecraft with two winged stages, both powered by
rocket engines, with horizontal takeoff and landing. After it had been
cancelled, a programme to construct a reusable launch vehicle (RLV)

Figure 2.4. Artist's impression of the Interim Hotol spaceplane
separating from the Antonov AN 225 aircraft (J.K. Davies, *Space
Exploration*, Chambers, Edinburgh, 1992).

as a substitute for the Space Shuttle was started. It included a number of technological demonstrators, the Delta Clipper (DC-X and then DC-XA), the X-34, the X-33, the X-37 and the Hyper-X, finally to the design of a true orbital vehicle, provisionally designated as *Venture Star*.

The Delta Clipper (Figure 2.5), a sort of one-third-scale working model of an unmanned orbital vehicle, was successfully flight tested in 1994 and 1995, although one of the prototypes was lost in an accident. It is a wingless single-stage rocket with vertical takeoff and landing. On re-entering the Earth's atmosphere the vehicle is controlled by the aerodynamic forces acting on its body, behaving as a *lifting body*. The full-scale orbital version should enter the atmosphere nose first and, at an altitude of about 20 km, rotate into the vertical position, nose upwards, and land using the thrust of its main engine. Some consider that this operating procedure reminds them of the science fiction cartoons of the 1930s. It does not even need a runway. The DC-XA is 13 m tall and has four liquid hydrogen–liquid oxygen engines, each capable of a thrust of 6,700 kg.

The DC-X (Delta Clipper – experimental) was built by McDonnell-Douglas for just less than US $60 million of funding from the Strategic Defense Initiative Organisation (SDIO), in only 24 months. Such a project is considered as a symbol of that non-bureaucratic approach to space exploration which many advocate.

The X-33, derived from the NASP (National Aerospace Plane) and X-30 programmes, is a true vertical-takeoff spaceplane. Although not capable of reaching low Earth orbit, it will reach a speed of Mach 15. Many different configurations were considered and, in the end, a lifting body with two short wings at the rear was chosen. It is 20.7 m wide, 20.4 m long, has a mass of 123 tonnes and will be powered by a liquid hydrogen–liquid oxygen rocket, without using air-breathing engines. The programme was cancelled in March 2001, before the first flight took place.

The X-34, the X-37 and the Hyper-X are smaller demonstrators. They are aimed at studying a propulsion system based on the

Figure 2.5. The technology demonstrator Delta Clipper DC-X (NASA photo).

Figure 2.6. Artist's impression of the reusable SSTO vehicle *Venture Star* (NASA image).

very innovative 'aerospike' engines, which adapt automatically to the ambient pressure and so are always very efficient (Hyper-X), novel structures and avionics (X-34), and the new technologies required for orbital vehicles (X-37, X-40A). All these craft will be unmanned research vehicles. The operational system which could replace the Space Shuttle, for manned and unmanned missions, may be the Venture Star (Figure 2.6). The present design, which is a very preliminary one, is an enlarged and more powerful version of the X-33, able to fly at up to Mach 30 and to carry 18 tonnes to low Earth orbit. NASA also cancelled the X-34 programme in March 2001. NASA's strategy is now to upgrade some of the Space Shuttle's systems so that it can continue launching payloads into space until at least 2012.

All these programmes will be managed by private industry, not only in the construction phase but – above all – in the operational phase. Many other privately financed programmes, often even more innovative than those mentioned above, are being suggested by companies created solely for that purpose. Perhaps the most innovative of these is the Roton, proposed by Rotary Rockets, consisting of a crewed SSTO vehicle which should take off vertically under the thrust of rockets located at the periphery of a 7 m disc, rotating at 720 rpm; the centrifugal force due to rotation means that the turbopumps usually needed to supply the propellants to the combustion chambers would be omitted. The vehicle would also land vertically, using a set of folding helicopter blades which are deployed after re-entry. Its proponents hold that it could soon make its maiden flight with the aim of reducing the launch cost (compared with that of the Space Shuttle) by a factor of ten, i.e. by an order of magnitude.

Often, in the history of technology, and especially in aviation, prizes offered by individuals or even by governments for the achievement of specific goals have been very effective in promoting progress. Examples range from the £20,000 prize offered in the 1700s by the British Parliament for the development of accurate instruments for sea navigation to the US $25,000 Orteig prize won by Charles Lindbergh with his non-stop New York–Paris flight in 1927. These prizes, and many others, triggered private investments much larger than the prize itself.

In 1996, Peter Diamandis and some businessmen from St. Louis, Missouri, offered the US $10 million X-Prize to be awarded to the first team that develops and flies a spaceship capable of launching three passengers to an altitude of 100 km on two consecutive flights within a two-week period. By the year 2000, 14 teams had entered the competition; by early 2002, none had achieved this feat.

A spaceplane such as the *Venture Star* could be engaged in sub-orbital flights, on routes between the United States and Japan, for instance, which should have a very large traffic volume in the future; the first project of this type, the National Aerospace Plane (NASP), had been aptly named *Orient Express*. The flight time between the United States and Japan would be of about one hour, while that between Europe and the East coast of the United States would be about 25–30 minutes.

The concept of a SOSS-OA-RRL (SubOrbital Single Stage – Once Around – Reusable Rocket Launcher) spaceplane, i.e. a rocket able to reach Mach 10 or 15 in the initial high acceleration phase, perform a suborbital flight and then reenter the atmosphere to any point on the ground, is also particularly interesting. Most of the propellant is used up in the initial high acceleration phase, while the cruise phase takes place outside the atmosphere. There is no air drag, and the weight of the vehicle is partially compensated for by the centrifugal acceleration along the trajectory rather than by aerodynamic lift, which causes drag. Its energy efficiency is far greater than that of a supersonic aircraft and the environmental advantages are significant: the rocket exhausts do not contain nitrogen oxides and the vehicle does not affect the delicate equilibrium of the middle and upper atmosphere, as the flight takes place mostly above 100 km altitude. The vehicle does not produce any sonic boom, not even in the final glide phase; it flies at supersonic speed only at very high altitude.

Spaceplanes of this kind, able to work in the suborbital mode and, with reduced payload, also to reach LEO, would be a fast, relatively economic and ecologically acceptable means of transportation. They would have, at least at the beginning, mainly a non-orbital use, and this should reduce the costs considerably, owing to the larger scale of construction and operation.

G.H. Stine in 'Comes the revolution'[10] analysed the economic consequences of the construction of a spaceplane, not forgetting to touch upon the legal issues and the air traffic control problems. The goal is that of lowering the cost to US $3,000 per kilogram of payload in low Earth orbit, about one tenth of the present cost of the Space Shuttle. The study starts from the obvious consideration that it will only be possible to obtain the support of business operators, and first of all of banks which will provide the capital, if the operation of a fleet of such spaceplanes becomes profitable. He shows that this condition

[10] G.H. Stine, Comes the revolution, in *Islands in the Sky*, Wiley, New York, 1996.

can be met in the near future, as soon as the vehicle is available, if the cost of each spaceplane is less than US $2 billion. Above all, the time needed to get ready for a mission must be less than 72 hours, with an aeronautics-style maintenance schedule and a ground crew of 50 per machine. The fuel and other costs are not expected to be a critical issue.

The availability of low-cost orbital transportation is predicted to create a large market.[11] The industrial and commercial activities linked with space will quickly multiply. Also, scientific activity will benefit greatly from this situation: today's trend is for research groups to buy commercially available equipment and services. In the future, scientists will be able to perform research in space at a fraction of the present cost. A fleet of spaceplanes will be the true gateway to space.

It is likely that a new spaceplane of some type will become operational in the not too distant future. It will need to reduce satellite launch costs by an order of magnitude and provide a transportation capability far larger than that available at present. A fleet of such spaceplanes will be able to satisfy all our requirements for transportation to and from space for many decades.

All new systems should have a very low environmental impact and, possibly, not use non-renewable resources. Spaceplanes powered by hydrogen–oxygen rockets or by liquid hydrogen ramjets dump water vapour in the atmosphere, with a very low environmental impact, except, perhaps, for the generation of clouds. If propellants are obtained from water, which can be decomposed into oxygen and hydrogen by electrolysis, only energy is required. The water produced by the combustion remains mostly in the atmosphere, with small quantities being lost into space.

Propulsion systems all require energy, which can be produced by any primary source, possibly a renewable source. The cost of the optimum launch method chosen will depend on the operating costs (capital, maintenance, etc.) and on the cost of energy. Ultimately, the

[11] See M. Rycroft (editor), *The Space Transportation Market: Evolution or Revolution?*, Kluwer, Dordrecht, 2000.

possibility of launching very many payloads into space will depend on the availability of energy at a reasonable price.

NON-REUSABLE ROCKETS

While waiting for spaceplanes, a large market for non-reusable launch vehicles still exists. It will probably exist for a long time, as reusable launchers will not be able to satisfy all transportation needs. It is possible that safety regulations will, for instance, prevent the transportation of liquid propellant rockets, needed to launch probes toward deep space, on spaceplanes. There will always be one-piece objects too large to fit into the cargo bay of a spaceplane, or military devices which military agencies will not entrust to commercial transportation operators. Another reason to continue with non-reusable rockets is to avoid the mistake of entrusting everything to a single class of machines. An accident to one spaceplane could ground all the others, blocking all activities as happened after the *Challenger* disaster.

It is unlikely that missions aimed at landing astronauts on the Moon or on Mars can be performed using reusable launchers. If the difficulties of assembling large spacecraft and, in particular, of transferring large quantities of cryogenic propellants from one spacecraft to another while in Earth orbit are to be avoided, a vehicle capable of carrying at least 100–120 tonnes to low Earth orbit is needed. Perhaps it is not impossible to build a reusable vehicle of that class, but the cost would be forbidding, particularly because the number of missions which such a large spaceplane would perform is very limited.

On the contrary, it is not impossible to meet this requirement with a non-reusable rocket, as both Saturn V and Energia are heavy-lift rockets of the required class. But only 18 Saturn V rockets were produced almost 30 years ago, and Energia is not operational due to the financial problems of the Russian Space Agency. The possibilities for such ambitious missions, apart from re-building Saturn V, which is not a viable solution, are to design a non-reusable rocket derived from the Space Shuttle (Figure 2.7), to resume the Energia programme or

Figure 2.7. Artist's impression of a Shuttle-derivative heavy-lift non-reusable rocket considered in the NASA Reference Mission for Mars (see Chapter 6; NASA image).

to develop a new launch vehicle which would, however, considerably add to the cost of the mission.

Since the assembly lines of the large rockets were closed down in the United States, the largest launch vehicle has been the Russian Energia. That, however, has flown only twice, at the end of the 1980s. The American decision was probably a mistake, also for the loss of know-how which followed. If a rocket like the Saturn V had to be built today, a long research and development programme would be needed, and not only to introduce all the updates due to the advances of the three most recent decades. Technological know-how in all fields cannot be considered as something acquired once and for ever. If some capabilities are not used for a certain time, they fade and can be regained only with large investments, in terms of human and financial resources.

Figure 2.8. An Ariane IV launch from Kourou, French Guyana (Arianespace photo).

After the *Challenger* disaster the construction of the Titan IV, a launcher able to carry slightly more than 20 tonnes into low Earth orbit, was resumed in the United States. The Proton, the largest of the fully operative Russian rockets, is of the same class.

Europe is in a good position in the market of non-reusable rockets with the Ariane V, able to carry about 20 tonnes into low Earth orbit and 6.8 tonnes into a geosynchronous orbit. Even earlier, thanks to Ariane IV (Figure 2.8), ESA was well placed for commercial launches, as well as providing launch services for all European scientific missions. Ariane V could have a module, the Crew Transfer Module, to carry people to the *International Space Station*. However, no definite schedule for the Crew Transfer Module is available.

Other countries – China, India, Japan and Spain – have developed launch vehicles and, besides launching their own satellites, compete in the international marketplace, often with mixed fortunes. Italy participates in European programmes and is also developing its own non-reusable rocket, Vega, with which it could compete in the market for the launch of small scientific and commercial satellites with a mass between 250 kg and 1,000 kg.

In a class of its own is the American Pegasus, a very small rocket launched from an aircraft flying at an altitude of about 13 km. It can put small satellites (400 kg) into low Earth orbit, or even send a small probe to the Moon.

Many experts hold that expendable launchers are intrinsically more convenient than reusable ones, even from an economic viewpoint. First, reusable systems need a large ground crew for maintenance and refitting the launch vehicle. The size of this crew does not depend on the number of launches which each vehicle performs each year. If this number is small, as in the case of the Space Shuttle, this cost is overwhelming. Expendable launchers do not need to carry into orbit the mass of the re-entry devices – the dry weight of the orbiter of the Space Shuttle is about 80 tonnes (roughly 50 tonnes for the wings, thermal protection and other re-entry and landing devices). This means that to carry a payload of 20 tonnes, at every launch 100 tonnes are put into orbit. Many expendable systems, designed for short periods of operation, can be lighter and cheaper than those which have to operate in space many times. Moreover, the greater scale of production allows a further cost reduction. There is much engineering work going on with the aim of simplifying the main components of expendable rocket engines and very important results have been achieved recently. A new nozzle designed for the Fastrac engine should cost about US $100,000, while a comparable component built using conventional technology would cost about US $800,000. The new R-68 engine, while having a thrust 50% higher than that of the Space Shuttle main engine, will have about one tenth the number of parts of the earlier design, and the designers hope for a cost reduction of one order of magnitude.

The treaties by which many Intercontinental Ballistic Missiles (ICBM) built during the Cold War must be destroyed open up new launch possibilities. The ICBMs can be converted to launching small payloads into orbit at a fraction of the cost of building a new rocket – even cheaper can be the use of submarine-based ballistic missiles. An example of the latter is the suborbital flight of the solar sail demonstrator built by the Planetary Society, which was launched by a Russian submarine in the Barents Sea (see Chapter 4). Submarine-based missiles can only perform suborbital flights, but there are projects under way to enable them to put small payloads into orbit.

However, the use of converted ballistic missiles has one severe drawback. Their reliability, particularly in the case of Russian missiles, is not very high – the philosophy was to launch many warheads, without bothering too much about the possibility of losing some of them. This is clearly not valid for the commercial satellite market, as the possibility of losing a payload and the need to find a replacement satellite makes the cost of this approach unacceptable.

SPACEPORTS

At present, spaceports, i.e. bases from which satellites are launched, are very few in number. The larger ones are those at Cape Canaveral, in Florida, USA, at Kourou, in Guyana, South America, at Baikonour, in Kazakhstan, and at Jiukuan, Xichang and Taiyuan, in China. To them the spaceport of Woomera, in Australia, and that at the Vandenberg Air Force Base, in California, must be added. All those are run by space agencies or by other governmental organisations.

With the growth of commercial space traffic, such bases are becoming insufficient – privately run spaceports are now operating and planned for the future. The first company in the United States to obtain the authorisation required to operate a spaceport (in September 1996) has been Spaceport Systems International, based near the Vandenberg Air Force Base. Among the other companies interested in performing commercial launches is Sea Launch, owned by Boeing together with some Russian, Ukranian and Norwegian companies. The rocket uses a two-stage Ukranian rocket with a Russian third stage. It can put about 5 tonnes into geostationary orbit (GEO). Sea Launch has recently completed a mobile launch base, obtained by modifying an oil drilling sea platform, and owns a 34,000 tonnes support ship. The system will operate from Long Beach, in California, where the customer will deliver the satellite to be integrated onto the rocket supplied by the company. The platform and the support ship will sail to a location such as that near Kiritimati, about 1,400 miles South of Hawaii, where the launch will take place. The system is particularly suitable for geosynchronous satellites, for which a launching site near the equator is beneficial. In the past, a platform at sea had been used with success by Italy for launching satellites, at the San Marco base, off the coast of Kenya, East Africa.

One of the ideas which motivated the development of many horizontal takeoff and landing reusable space vehicles, like the German *Sänger*, was that of operating from conventional airports. Nowadays, it seems that this goal will be postponed for some tens of years. Thus spaceports, whether run by governmental or private organisations, will remain for a long time as very specialised and rare facilities.

GUNS, SKYHOOKS AND SPACE FOUNTAINS

It is likely that spaceplanes and rockets will provide transportation between the surface of Earth and orbiting space stations even if the volume of traffic becomes quite large. Moreover, spaceplanes promise to have low operating costs. This notwithstanding, many possible ways to launch a payload into space have been studied. In some cases the ideas which were put forward seemed weird and impossible; in other cases they were premature. Sometimes the obstacles are just practical ones, such as the lack of suitable materials or of sufficiently compact energy sources. In other cases they are theoretical, due to the lack of the basic knowledge required to understand and exploit some physical phenomena.

It is interesting to consider some of these extreme ideas, not because they will be practically important in the near, or even distant, future, but to illustrate some possibilities which cannot be ruled out. The main research lines are:

- Guns and electromagnetic launchers
- Skyhooks
- Laser or microwave vehicles
- Space fountains.

Guns and electromagnetic launchers

Jules Verne's idea of using a gun to launch a spacecraft towards the Moon has been considered as being impossible in the previous sections. However, a device fixed to the surface of a planet can be used to impart an initial velocity to a spacecraft, at least saving on the first stage or on the boosters. Simple computations show that, using a 'gun' with a barrel 25 km long and applying an acceleration equal to 50 times the gravitational acceleration, 50g, it is possible to reach 5,000 m/s, i.e. 18,000 km/h. This is about two thirds of the first cosmic velocity. A gun barrel 25 km long might appear to be an absurd idea, but it could be built by digging an inclined tunnel in a mountain, with a technology similar to that used for making the tunnels which cross the Alps. An acceleration 50 times the acceleration of gravity is very large, beyond the allowable limits for human beings, but could be applied to suitably designed robotic payloads.

The spacecraft could be launched using a charge, as in artillery. More probably a linear motor could be combined with a magnetic levitation device, in a way not dissimilar to maglev trains. A very large power source would be needed for the launch, and some energy storage device is also required. Several projects of this type have been proposed and many papers have been presented at magnetic levitation symposia. It was also said that the designer of the enormous gun which Saddam Hussein may have tried to build at the end of the Gulf War had it in mind to build such a system to launch satellites.

Even if they are not to be ruled out, devices of this type are not straightforward and, particularly if a spaceplane is built, they may not be economically practical either. A limited version in which the electromagnetic launcher imparts some initial velocity to a spaceplane is more feasible, and both NASA and the Russian space agency are working on this. Because in any space launch a huge amount of propellant is used in the first few seconds of the flight, to provide the thrust needed not only to support the weight of the rocket but also to build up the first few meters per second of speed, large fuel savings

Figure 2.9. Artist's impression of an electromagnetic launcher on the Moon. On the left are the solar panel arrays (NASA drawing).

could be obtained if the initial acceleration, even if only to a speed of some 200 m/s, could be performed by an electromagnetic launcher.

Electromagnetic propulsion could also be convenient for launches from celestial bodies smaller than the Earth. They have been suggested for launching objects, and perhaps manned spacecraft, towards the Earth from a lunar base (Figure 2.9). The gravitational acceleration on the surface of the Moon is so low that to send an object to the Earth a velocity of only 8,400 km/hr is needed. This can be reached with an acceleration of only 10g in about 27 km. If an acceleration of 100g could be accepted, the length of the device could be reduced to only 2.7 km and, using an electromagnetic launcher, a tunnel would not even be necessary – a magnetic rail on the side of a mountain would suffice. The electric power needed for the launch is not enormously large. It is about 3,000 kW per kilogram of material launched to the Earth, with an acceleration of 100g, assuming an overall efficiency of 75%. Also, the amounts of energy are very reasonable, about 1 kWh to send every kilogram, computed using the same value of the efficiency. A square with sides of 180 m covered with solar cells would be sufficient to power a device for launching a pellet of 1 kg every few seconds from the Moon to the Earth, i.e. sending some tens of tonnes every day.

Systems of the same type could be even more competitive for delivering material mined on the asteroids toward the Earth. The very weak gravitational fields of the asteroids mean that much less power is required, keeping in mind the need to send the material to a collecting point on (or near) the Earth and not only to put it clear of the asteroid.

Electromagnetic launchers have been described in detail by Gerard O'Neill,[12] who suggested that electromagnetic launchers could be used 'the other way around' as rockets to allow space habitats or other spacecraft to change their orbits. The launcher is attached to a large spacecraft, and the pellets which it shoots into space are the reaction mass, like the gas jet in conventional rockets.

Skyhooks and tethers

A device which has been proposed several times is the so-called 'skyhook'.[13] Imagine deploying a cable downwards from a satellite in a geosynchronous equatorial orbit; for the satellite to remain in the same orbit, a cable must be deployed in the upward direction too. The first cable is so long that it reaches the surface of the planet at a point on its equator. As the satellite remains vertically above that point on the surface of Earth, the cable will be a vertical column pointing to the sky and a lift leading to the geosynchronous orbit can thus be built. As an added bonus, the other cable, pointing toward deep space, can be used to launch outbound vehicles.

[12] G.K. O'Neill, *The High Frontier: Human Colonies in Space*, Bantam Books, New York, 1977.
[13] A detailed description of such systems can be found in R.L. Forward, *Indistinguishable from Magic*, Baen, Riverdale, New York, 1995, and in R.M. Zubrin, The hypersonic skyhook, in *Islands in the Sky*, Wiley, New York, 1996. The term skyhook used in this context must not be confused with the 'skyhook dampers' used in automotive active suspensions.

It is simple and seems to be cheap: unfortunately, its implementation is very difficult and controversial. Until a few years ago most specialists thought that it could not work, not for technological reasons, but for conceptual ones, at least in the case of a massive planet like the Earth. First, the cable must be 36,000 km long, but this is not the point. To carry its own weight it would be so stressed that even the strongest materials would break. Worse still, the stress is larger than the theoretical strength of ideal materials. It is true that the cross section of the cable can be made larger and larger going upwards to reduce the stresses, but soon its cross section would become so large that enormous quantities of material would be needed. So space elevators appear to belong only to science fiction, as in the novel *Fountains of Paradise* by Arthur C. Clarke.

Recently, however, these opinions have been reviewed. Microtechnology allows the construction of materials, like carbon nanotubes,[14] with a strength to weight ratio far higher than yesterday's wildest dreams; a serious NASA study is considering the construction of space elevators (Figure 2.10). Even if such a device may be a technological possibility, it is economically justified only for very high volumes of traffic between the planet and space, owing to its huge investment and maintenance costs. The cost of sending each payload into Earth orbit or into outer space could become quite low – if the inbound traffic matches the outbound traffic, the net quantity of energy involved might well be almost zero since the payloads going down would supply the energy needed by those going up. The most optimistic forecasts suggest that space elevators might be built – if ever – in the second half of the twenty-first century.

The starting point of the space elevator must be on the Earth's equator. In the novel by Arthur C. Clarke it was a hypothetical island south of Ceylon, but perhaps a higher place would be a better

---

[14] Carbon nanotubes are tubular carbon filaments with a wall thickness of a single atom (single-walled nanotubes) or a few atoms (multi-walled nanotubes) without a true chemical bond in the radial direction. They have outstanding mechanical properties (strength and stiffness) and very low electrical resistance.

Figure 2.10. A space elevator, seen from a geosynchronous satellite, looking down towards the Earth (NASA image).

choice – somewhere North of Quito (Ecuador) or East of Nanyuki (Kenya) are among the few places on the equator at an altitude of several thousand metres. In this way a few kilometres of cable could be saved.

More limited applications of the concept are feasible. A tether can be used to move objects upwards or downwards from a satellite. This has been experimentally demonstrated by the Tethered System Satellite (TSS, Figure 2.11). Unfortunately, the cable broke during the 1996 experiment when a small satellite was about 20 km from

Figure 2.11. Artist's impression of the Tethered System Satellite (NASA image).

the Space Shuttle, with the cable deployed almost to its full length. The experiment was a success, in that it demonstrated that the three-body system comprised of the Space Shuttle, the satellite and the cable connecting them is dynamically stable. If the tether is made of a conducting material, it can generate electricity, as in its motion it cuts lines of force of the Earth's magnetic field. This was also demonstrated practically in this experiment. Obviously, no electric power is generated for free: if a current is circulating through the circuit and power is extracted from the system, the system loses energy and the orbit slowly decays. This tethered system was proposed by Giuseppe Colombo, but unfortunately he did not live long enough to see it operating. But, in his honour, ESA's mission to Mercury is called *BepiColombo*.

Apart from the generation of electric power, which could be useful for a space station, a tethered system has many applications. If the tether is deployed downwards, its speed is lower than the orbital speed. Conversely, the end of the tether deployed upwards has a speed higher than that needed to remain in orbit. A tethered system permanently positioned under the space station allows any object which is lowered along the cable and then released to re-enter without using retro-rockets to deorbit it. This could be a very economical way to send objects back to the Earth from the space station. Rotating tethers have been suggested for bringing objects from one orbit to another, either lower or higher, and even for inserting a spacecraft onto a trajectory to the Moon from an orbit around the Earth.

Extending the concept of tethers, the *hypersonic skyhook* is obtained. A cable, far longer than the 20 km of the TSS experiment, is lowered into the middle atmosphere from a satellite in low Earth orbit. As the satellite is not synchronous, the end of the cable is seen by any point on the Earth as moving at a very high hypersonic speed just under the satellite.[15] Hypersonic aircraft could then carry objects up to the end of the cable, which would act as a relay station. From that point on, the payload could be raised along the cable, as if it were in a lift. The main advantage would be that hypersonic aircraft with air-breathing engines could be used. The downward cable also has to be accompanied by an upward one, which could be used to launch payloads deeper into space.

A system of this type is not theoretically impossible, but it seems that it cannot lead to interesting applications in the near term. Skyhooks, hypersonic or not, appear to be more curiosities or possibilities for the distant future. On the other hand, more traditional tethered space systems could have important applications in the near future.

---

[15] It cannot go down to the surface of the Earth because it would move across it at a speed of several thousand kilometres per hour. It could not even reach the lower atmosphere, where the air drag would cause it to break up.

Laser and microwave vehicles

Systems in which the energy needed for motion is supplied to the vehicles from the outside during motion are very common in ground transportation (trains, trolleybuses, etc.). If the same thing were possible also in the case of air or space vehicles, large weight savings could be achieved, together with an increase of performance and an overall cost reduction. The vehicle must carry on board the fluid needed to produce a jet, but the energy needed to accelerate it can be supplied from the outside. For space vehicles lifting off from a planet with an atmosphere, however, the jet can be formed using the gases taken from the atmosphere, e.g. by an air-breathing engine.

Energy can be generated on the ground or in space, and then transmitted to a satellite using laser or microwave beams. In the former case the beam can be focused into a ring which heats air to a temperature of some tens of thousands of degrees, causing it to expand and to form an high velocity jet. Using a 10 kW carbon dioxide laser pulsed at 28 Hz, spin-stabilised lightweight discs of 10 to 15 cm diameter have been pushed to altitudes of about 30 m in 3 seconds. Far higher powers are required for actual space applications, but the proponents of the concept intend to demonstrate it by orbiting satellites of up to 100 kg by using a 100 MW laser beam. As the laser is pulsed, it is not too difficult to reach the values of power required. Very powerful lasers are being developed for military purposes.

Although microwave sources cannot reach the same high power densities as lasers, they are less costly and can be scaled up more readily. A microwave beam can be focused to a point ahead of a space vehicle to produce a shock wave in front of it. The spacecraft is then in a zone where its relative velocity can be low even if the vehicle travels fast, reducing drag and aerodynamic heating. The microwave beam supplies energy to the engine which breathes in air from a zone behind the shock wave and ejects it at a higher speed behind the vehicle. The thrust so obtained can accelerate the vehicle up to such a speed that it orbits the Earth. Some hypersonic wind tunnel experiments have demonstrated the technical feasibility of such a device at a speed of Mach 10.

Although working mostly on microwave driven sails (see Chapter 4), Microwave Sciences Inc., a California based firm, recently used a 25 kW X-band (7.16 GHz) transmitter to levitate in the Earth's gravity and accelerate upwards a light sail made of carbon microtruss fabric. This is another material made possible by micro- and nano-technology, like carbon nanotubes.

The possibility of using microwave beams to transfer energy from the ground has also been verified in Japan and Canada using radio controlled aeroplanes, with electric motors driving their propellers.

Space fountains

The concept of a *space fountain* has been discussed and its feasibility demonstrated, at least from a strictly technological point of view.[16] A *space fountain* is an engineering feat which requires huge invest-ments, far greater than those involved in the tunnel under the English Channel and perhaps even in space elevators. It can be only justified if the volume of traffic between the surface of the Earth and low Earth orbit is very large indeed.

The concept is easily understood: many pellets are launched ver-tically upwards one after the other. They rise to a certain altitude and then start falling down, as do the drops of water of the fountain after which the device is named. When the pellets are back at the altitude from which they started, they enter a curved tunnel to bend their tra-jectory upwards again. They are then launched for a new run, as the water of a fountain which is pumped over and over again, to form a continuous water column. Once the device is started and the pellets are set in motion the only energy required is that needed for over-coming friction and other losses, as on the downward leg the pellets recover the speed which they lost in the upward motion.

---

[16] R.L. Forward, *Indistinguishable from Magic*, Baen, Riverdale, New York, 1995.

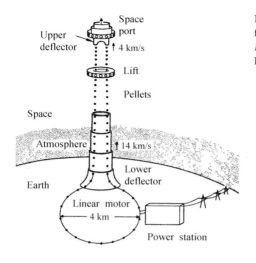

Figure 2.12. Sketch of a space fountain (R. L. Forward, *Indistinguishable from Magic*, Baen, New York, 1995).

If a deflector is located on the top of the column of pellets at an altitude below that at which they stop, the deflector receives an upward force which balances its weight, in a way which is similar to the table-tennis ball which is kept in equilibrium on top of a water jet. To complete the structure it is possible to add a casing around the pellet stream. This can be made in several sections, each one of them being directly supported by the pellets: a linear electric motor brakes the pellets on the upward leg and accelerates those going down. By doing so each section receives a force which balances its weight. The result is a sort of tower, which can be hundreds or thousands of kilometres tall, going out of the atmosphere (Figure 2.12). The pellets could also be launched in an inclined direction, forming space arches and structures of different kinds.

In the figure a speed of 14 km/s, or 50,000 km/hr, is indicated for the pellets; although this is higher than the escape velocity from the Earth, the pellets are not lost in space as they are slowed down in their upward trajectory (and accelerated in the downward part) to balance the weight of the tower. It is an active and controlled system, as power needs to be supplied to allow the tower to stand and all systems must be continuously controlled to achieve a stable working condition. A number of precautions must be taken to guarantee the correct working of all the vital systems. Such active systems are now entrusted with the control of oscillations, due to the wind or to an earthquake, in large buildings and bridges.

Small space fountains could be used in the future to build very high buildings or towers. It is not yet certain that spaceports of this type will actually be built. That will mainly depend on the cost of such systems compared with that of other, more conventional, rockets or spaceplanes.

As a final consideration, it is clear that large structures like skyhooks or space fountains may be very vulnerable to terrorist attacks or to acts of war. The consequences of the collapse of a skyhook would be awful: if the cable is severed close to the satellite, it would fall down, winding almost all the way around the Earth at the equator. A collapsing space fountain would damage a large area, not just the area in the immediate neighbourhood of its base.

3   Cities and factories in space?

American Astronauts carried out many experiments in space aboard *Skylab* during the early 1970s. Soviet and Russian cosmonauts have since clocked up many more hours aboard *Salyut* space stations and the *Mir* space station. Now the *International Space Station* is being constructed and beginning to be used, not only as a laboratory but also as a prototyping factory. Larger space habitats have been proposed. Living and working in space is hazardous, what with radiation and space debris. However, space is a good place to perform scientific and technological research and to generate energy, certainly that needed for each specific mission and perhaps also to beam it down to the Earth.

ORBITAL LABS AND EARLIER SPACE STATIONS

Space stations have been one of the favourite subjects of science fiction stories, from *The Brick Moon* by E.E. Hale (1869), to *Island in the Sky* by A.C. Clarke (1952) and to the *Deep Space 9* TV series. All space pioneers, starting with Tsiolkovsky, tried their hand at designing them. The need to supply some form of centrifugal acceleration to substitute for gravitation was clear from the earliest projects and the living quarters at least were usually given some rotational motion. Perhaps the most famous was the large space station in the shape of a wheel designed by Wernher Von Braun in the early 1950s (Figure 1.3). It had a number of uses, both civil and military, including acting as a spaceport and fulfilling the tasks of modern meteorological and Earth-observation satellites.

Since the 1960s a space station has been included in the medium-term plans of both NASA and the Russian Space Agency. But, as time progressed, the projects were reduced and simplified. The idea of building a large, rotating space station has been postponed until the cost of attaining satellite velocities is much reduced. The space agencies concentrated their efforts on smaller orbiting laboratories. At the end of the 1960s, the Americans had at their disposal the hardware developed for the *Apollo* programme, which included the largest rocket ever to become operational (the Energia rocket cannot be considered as such). The idea of using the large liquid hydrogen tank of the third stage of a Saturn V rocket as the basis for a space station was proposed. The result was *Skylab* (Figure 3.1), the first large space laboratory. Its internal space was 15 m in length, had a diameter of 6.6 m, and a volume of 330 m^3. It was launched on May 14, 1973, and remained in orbit until July 11, 1979; three different crews, each of three men, lived in space for weeks at a time. The last *Skylab* mission lasted slightly more than 84 days.

The experiments performed on the *Skylab* were mainly concerned with space medicine and the adaptation of the human body to the space environment, but also many biology and materials science experiments were carried on board. A good deal of activity was also dedicated to astronomy, particularly to the study of the Sun. The Apollo Telescope Mount was a large module, with a mass of more than 10 tonnes, carrying a number of telescopes and having its own power and control units.

Even if the first large orbital laboratory was American, the greatest experience of life aboard space stations has been accumulated by the Russians, with their *Salyut* stations and then with the space station *Mir* (Figure 3.2). *Salyut* was the name given to a number of early space stations of different types, including the civilian stations initially designated as *Salyut* and the *Almaz* military ones. They were carried into orbit in their fully assembled form by Proton rockets. The *Mir* space station was far larger, consisting of several modules

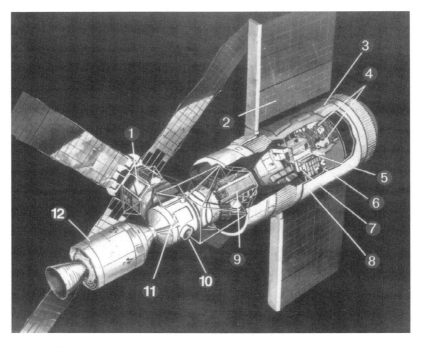

Figure 3.1. *Skylab.* 1: solar observatory (ATM, Apollo Telescope Mount); 2: solar panel; 3: shield for protection against meteorites; 4: experiments; 5: sleeping quarters; 6: eating area; 7: storage area; 8: laboratory; 9: airlock module; 10: secondary docking system; 11: main docking system; 12: *Apollo* spacecraft (NASA diagram).

(*Kvant-1*, *Kvant-2*, *Kristall*, *Spektr* and *Priroda*, plus a central module) launched at intervals between 1986 and 1996. Thanks to these Russian space stations, many astronauts from various countries, including a few Americans, could spend very long periods in space, in a microgravity environment. While the early Russian stations hosted only cosmonauts from the countries of the Communist block, *Mir* has also been home to Western astronauts. The European Space Agency had a joint programme with the Russians, named *Euromir*, and so did NASA.

Crew changes and the resupply of materials were performed through non-reusable launchers and small space capsules which re-enter the Earth's atmosphere using parachutes, as did the early

Figure 3.2. Photograph of the Russian space station *Mir*, taken from the Space Shuttle *Atlantis* (NASA photo).

American spacecraft. This demonstrates that winged shuttles or spaceplanes are not strictly needed to operate space stations. The Russians developed an automatic cargo spacecraft, *Progress*, to resupply space stations. However, a docking device compatible with that of the Space Shuttle has been developed and American astronauts (Figures 3.3 and 3.4) have reached the Russian space station via their own Space Transportation System.

The size of the *Mir* station was large enough for a few people (usually three) to live and work for extended periods of time, but its comfort has been greatly reduced by a general decay due to ageing. Apart from the usual problems of living in a confined space, one of

Figure 3.3. The Space Shuttle *Atlantis* docked to the space station *Mir* on April 9, 1996 (NASA photo).

Figure 3.4. The *Spacelab J* in the cargo bay of the Space Shuttle *Endeavour* (flight STS 47, September 12, 1992). Note particularly the thermal blankets and the passageway for the astronauts to reach the laboratory from the Shuttle's crew compartment (NASA photo).

the worst things which made life unpleasant there was infestation by a bad smelling mould.

The *Mir* space station became operational in 1986. It was planned that it should be replaced during the 1990s, but delays in the launches of the additional modules and severe budget cuts compelled the Russian Space Agency to extend its life.

During its long life,[1] the *Mir* station suffered accidents and malfunctions, which almost caused the emergency evacuation of the crew. The most serious one occurred when an automatic cargo ship *Progress* crashed against the solar cell panels during a manual docking manoeuvre. The crew was able to stabilise the spacecraft after the shock caused a loss of control and a power black-out, and during subsequent missions the worst consequences of the accident were repaired. These accidents caused much concern in the United States about the actual safety of this ageing space station and American participation aboard it soon ended.

Many useful tasks, from space medicine studies to experiments in many branches of technology, have been carried out aboard *Mir*. One of the most important results is the experience gained running a space station for more than 10 years. The cosmonaut who spent the longest time in space was the Russian medical doctor Valeri Polyakov, who heads the Space Medicine Centre of the Yuri Gagarin Centre, with fourteen months on the *Mir* station from January 1994 to March 1995. Polyakov had previously spent 8 months in space, in 1988, and so he held the world record for almost 2 years of spaceflight. That record was broken in 1999 by Sergei Avdeyev, who in total spent just more than 2 years in space, a year and 2 weeks of them aboard *Mir*.

Plans to deorbit *Mir* gathered momentum early in 2001 with the launch of a *Progress* cargo vehicle. But then *Mir*'s attitude control system suffered a hiccup, so that re-entry was delayed. On 21 March three bursts of the *Progress* engines took place, lowering *Mir* deeper into the Earth's atmosphere. First, the solar arrays broke off. People

[1] Details are discussed in B. Macnamara, *Into the Final Frontier: The Human Exploration of Space*, Harcourt, Orlando, Florida, 2001, and in B. Harvey, *Russia in Space: The Failed Frontier?*, Springer-Praxis, Berlin–Chichester, 2001.

living in Fiji on 23 March 2001 glimpsed several fiery modules before these plunged harmlessly into the ocean between New Zealand and Chile.

The European *Spacelab* module, whose pressurised living and working cylinders were built by the Italian space industry, is a sort of temporary space station (Figure 3.4). *Spacelab* cannot work in an autonomous way, being carried in the Space Shuttle cargo bay. It remains there for the duration of the mission, and many scientific and technological experiments have been performed there. It is not a true space station, as it comes back to the Earth after each mission, but the experiments carried out opened the way to the *International Space Station*. Rather similar is the space laboratory *Spacehab*, built by Alenia Spazio and operated by a private US company, which rents it out to some space agencies.

THE *INTERNATIONAL SPACE STATION*

After years of discussions, the design of the *International Space Station* (ISS) *Alpha* (Figure 1.12) has finally been defined. The method chosen was based on the in-orbit assembly of parts carried into space by the Space Shuttle or by other 20-tonne-class launchers (mostly the Russian Proton rocket). To discard the technically simpler, and perhaps more economical, method based on a few larger parts, more or less of the *Skylab* size (which would have required a heavy-lift launcher, so precious for human Moon or Mars exploration) was a political issue. Its main motivation was the usual one – NASA still needs payloads for the Space Shuttle.

Construction of the ISS in orbit started in 1998 with the launch of the first two modules – the Russian FGB (from the initials of Functional Cargo Block in Russian) *Zarya* (Figure 3.5) and the American *Unity Node 1*.

The decision to build a large space station caused many arguments in the scientific community, not only in America, which are not yet completely resolved. Many scientists were, and still are, against it, saying that it is almost useless for scientific work and will absorb a

Figure 3.5. The first module of the *International Space Station*, the FGB, *Zarya*.

huge quantity of resources which, if used for more productive science missions, could lead to more important scientific results being obtained. Many outstanding space scientists, such as James Van Allen, who worked on American space programmes from their inception, discovering the radiation belts of the Earth which are named after him, are among those who oppose the space station. Also Brian O'Leary, scientist and astronaut, was initially against its construction, even if later he changed his mind. If some space scientists with their experience express these doubts, they cannot be dismissed lightly.

The motion of the astronauts spoils the microgravity environment, and this is an important issue. The *Hubble Space Telescope* could never be installed on a space station, as the human presence would induce movements and so reduce the quality of the images. But it could not be installed on a free-flying satellite near a space station either, as the gases ejected by the rocket engines of space vehicles operating near the station would reduce its performance.

If space is solely regarded as a scientific laboratory, it is best to launch small unmanned spacecraft, each performing a particular experiment. But if space is to become the *final frontier*, human presence in space is essential. Then a space station not only makes sense, but it is a prerequisite.

The Space Station is important in many applied science experiments, mainly concerned with materials technology and biology. Other developments are more mundane – but likely to bring a quicker return. As an example, the Japanese firm Shiseido conducted experiments in space to develop a new perfume which is now being marketed. The Space Station is then, in a sense, a prototype for future space factories. It is a vital asset for research on space medicine and therefore is essential for the preparation for future long-range manned space missions. Many pieces of hardware which will be essential for these exploration expeditions will be tested on the ISS; an example is the innovative TransHab module, an inflatable pressurised habitat designed to house the astronauts on their way to Mars (Figure 3.6).

Figure 3.6. Artist's impression of the inflatable TransHab module being tested at the *International Space Station* (NASA image).

The crews which will leave the Earth for Mars will need long training periods in space, to learn how to live and work there, and to get used to the conditions in which they will live for months.

Astronauts will also have operational tasks to perform on the space station. They will service satellites, build large space structures which cannot be launched already assembled, such as extensions to the space station itself, or an orbiting power station, or launch spacecraft which will explore the solar system. These aspects – its use as a factory, depot and training centre – could become more important than the strictly scientific aspects of the International Space Station.

Pioneering space station studies foresaw a 'space tug', i.e. a reusable Orbital Transfer Vehicle (OTV), to put satellites into geostationary orbits. The OTV, planned as an upper stage of the Space Shuttle, was to be the European contribution to the Shuttle program. Later, a nuclear OTV, able to reach the Moon, was also planned, and many of the efforts in the Nuclear Engine for Rocket Vehicle Applications (NERVA) programme were made in this direction.

As time passed and space programmes were downsized, it was realised that this would lead to non-optimal conditions for all launches beyond LEO, since the orbit of the space station is not a good starting point for a wide variety of missions, and the idea was abandoned. Following the success of the *Hubble Space Telescope* repair and upgrade missions, the idea of an orbiting manned vehicle to maintain scientific and commercial satellites (even in GEO) is being considered again. When cheaper means of achieving low Earth orbits are available, the *International Space Station* may regain its role as a 'dockyard'.

The crew of the space station will mainly be composed of astronauts training for long-range missions, scientists and engineers.[2] However, many other people will visit it and, for short periods, journalists, educators and politicians will be sent up by space agencies to

[2] It was initially planned that the crew should number six or seven, which requires two emergency return vehicles. If there is only one, the maximum number of astronauts aboard the ISS at any one time will be three.

allow them to gain first-hand experience of life in space. NASA had a programme of this type with the Space Shuttle: two members of the US Congress have already made a flight into space and, among the people killed in the *Challenger* accident, there was a teacher, Christa MacAuliffe, chosen from among thousands of applicants. After that disaster the teacher programme was stopped, but it will probably be resumed in due course.

A wealthy American businessman, Dennis Tito, paid the Russians US $20 million for a seat on a Soyuz vehicle which took him to the ISS for a six-day visit in April and May 2001. By agreement, he spent most of his time in the Russian segment of the ISS, happily watching the world go by some 350 km below.

Thousands of people have already booked a tourist's stay in space through private agencies. There is no doubt that a decrease of the costs of launches, and the greater availability of transport to orbit and of space on the space station, will allow space tourism to start in earnest.

The *International Space Station* is a true international undertaking. Its modules and other components are built and launched by several national and multinational space agencies, such as the American, Russian and Canadian Space Agencies, ESA or NASDA. ESA, in particular, will participate with a pressurised module, *Columbus.* The Italian Space Agency (ASI, Agenzia Spaziale Italiana) is participating with its own pressurised modules, the Multipurpose Logistics Modules (MPLM) Leonardo, Raffaello and Donatello. Both modules demonstrate the firm commitment of the Italian space industry, mainly Alenia Spazio, who will exploit its experience gained in the *Spacelab* and *Spacehab* programmes.[3] The direct participation of the Italian Space Agency in the *International Space Station* will enable Italy to use the scientific equipment aboard the station from the very beginning, while other European nations will have to wait until the end of the construction phase.

---

[3] E. Vallerani, *L'Italia e lo Spazio*, McGraw-Hill Italia, Milan, 1995.

The task of refuelling the station will be entrusted to the Russians, with their *Progress* spacecraft. This is a critical issue, as the very tenuous atmosphere at the average altitude of 400 km above the Earth's surface would cause the orbit to decay in a few years. A few times a year (typically four times) small thrusters will be fired to raise the orbit. There is some concern about this, in case a worsening of the economic conditions in Russia prevents the necessary refuelling from taking place. If so, other partners would be forced to develop new hardware to perform this task. It is worth pointing out that the attitude control for the ISS will be performed using two sets of reaction wheels, located in both the American and Russian modules, and additional thrusters.

After the first launches in 1998, various other modules and structures are being added to the ISS over several years, and for a long time the crew will have the primary task of assembling the huge structure. This will be an example of international cooperation never previously attempted, and many difficulties will have to be overcome. The most trivial aspect is the compatibility of tens of joints between parts built in places thousands of kilometres apart, by firms with different traditions, standards and operating procedures, which will be put in contact with each other for the first time in low Earth orbit. Some 43 launches of the Space Shuttle and the Russian Proton rocket will be needed to carry everything into space.

Once finished, it could be manned by a team of up to six astronauts, living in an internal space similar in size to that of two large airliners of the Boeing 747 class. Its mass will be of 463 tonnes, with an overall length of 75 m, a width of 109 m and a height of 40 m.

EFFECTS OF MICROGRAVITY ON THE HUMAN BODY

Even if microgravity is a very interesting condition for many scientific experiments, it could be detrimental for all living organisms. Our human anatomy has evolved on the surface of the Earth in an environment with a well-determined value of gravitational acceleration. Any decrease (or, even more, any increase) of gravitational acceleration will affect the operation of many vital organs. Before Laika, the first living being to withstand microgravity conditions, survived for a fairly long time on the *Sputnik 2* satellite, some biologists held that life was utterly impossible without a gravitational field.

Now we know that humans (and animals) can survive for a very long time in conditions of weightlessness, but their health is affected. Some effects, like space sickness, a combination of nausea, sweating, vomiting and loss of appetite, occur in the first few days of a space mission. Other symptoms develop more gradually, but have more lasting consequences. There is a general redistribution of all bodily fluids, cardiovascular changes, loss of bone material, and a height increase. The human body is grossly overdesigned for conditions of weightlessness and, in an effort to compensate, reduces the superfluous parts – the bones, the muscles, the heart, and so on. These changes are of little consequence in orbit, but problematical for withstanding the stresses of re-entry and on returning to Earth.

Very long periods in space, as experienced by Russian cosmonauts aboard the *Mir* space station, show that such damage may be limited with regular physical exercise. After more than one year in orbit, re-adaptation to normal gravity conditions on Earth was fairly easy, if a proper exercise regime had been followed in space.

The only way to create artificial gravity in a space station is to rotate it. However, the centrifugal acceleration so created is not exactly equivalent to the gravitational acceleration and, particularly if the space vehicle is not very large, the feeling which the people on board would experience might be strange, at least at the beginning. This is because there is also a Coriolis acceleration, which is felt only when moving in a radial direction. It causes the astronaut to feel as if the vertical direction is inclined at a slight angle to the radial direction. Another effect of artificial gravity is that distant objects in the space station will look as if they are leaning forwards (see Figure 3.7). Human beings should become used to artificial gravity aboard a space station – it will be much better than weightlessness.

Figure 3.7. Inside view of a very large wheel-shaped rotating space station; note that distant objects appear to be leaning towards the viewer (NASA image).

Thus, we know much about the effects on humans of a gravitational acceleration with a value[4] of $1g$ or near zero. But we do not know what is the minimum value of acceleration which removes the most inconvenient effects of weightlessness. The fact that the astronauts who walked and worked in $0.17g$ on the Moon adapted themselves so easily lets us hope that such a value of acceleration is far better than zero, and that it will be not necessary to produce an acceleration close to $1g$ in space stations. This is important, since the lower the gravitational acceleration the less stressed will be the structure and so the easier it will be to build. Also, the Coriolis acceleration will not be too large.

[4] An acceleration of $1g$ is the acceleration equal to the mean gravitational acceleration on the surface of the Earth (9.807 m/s^2).

Conditions of weightlessness in space might be beneficial for people suffering from heart problems, particularly in old age, or when waiting for a heart transplant. But before being able to use this we must develop launch systems without strong accelerations – a space lift would be the best solution. The space flight of John Glenn late in 1998 showed that old people can reach Earth orbit, but he was an exceptionally healthy – and well trained – specimen.

RADIATION AND SPACE DEBRIS

Once in low Earth orbit (LEO), human beings must face two dangers, namely radiation and objects of all sizes which might hit the spacecraft at high speed.

The region of space in which low Earth orbits lie is generally protected from radiation from outer space and solar flares by the Earth's magnetosphere (Figure 1.19). If the orbit lies below the Van Allen radiation belts (Figure 3.8), the radiation intensity is not high (except, due to a quirk of the Earth's magnetic field, off the coast of Brazil); the amount of shielding to protect people and sensitive

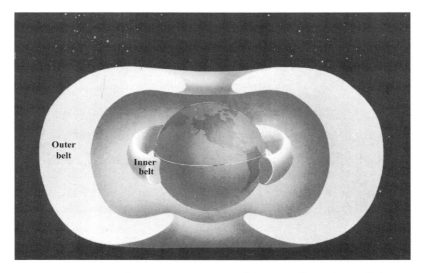

Figure 3.8. Schematic cross-section of the Van Allen radiation belts.

electronic equipment is minimal. Even in the event of solar flares, the total amount of radiation which astronauts would receive is not too dangerous in LEO.[5] Radiation shielding would be required in a space station if nuclear reactors or engines, or radioisotope thermo-electric generators, were present.

The dangers linked with freely flying objects – whether natural or man-made – are real in any orbit. In LEO altitudes between 500 and 1,000 km, the situation is worse than in higher orbits, which are less affected by artificially produced space debris.

Humankind is 'polluting' space close to the Earth, with all sorts of objects, spanning the size range from dust, paint flakes, lost screwdrivers to decommissioned spacecraft or rocket shells which remain in orbit. The number of these objects is not dangerous in itself, since space is already populated by natural objects – from micrometeoroids to remnants of comets to true large asteroids. The point is that artificial debris tends to be concentrated along the few orbits which are most used, and in these orbits they represent an ever-present danger.[6]

The large pieces of space debris are accurately tracked using radars and telescopes; their number and positions are well known. In 1996, there were about 4,000 detectable objects, and now there are some 8,500 objects with sizes about 10 cm or greater. Although new debris is always being produced, the older debris decays owing to the drag (friction) with the upper atmosphere – the objects re-enter the atmosphere, being completely destroyed before reaching the ground. Only metre-sized objects have some chance of reaching the ground, and so constituting a danger for people, a danger which is much lower than that due to natural objects. The 11-year solar cycle has a strong effect on both satellites in LEO and space debris, since the density of

---

[5] Space Studies Board, *Radiation and the International Space Station: Recommendations to Reduce Risk*, National Academic Press, Washington, DC, 2000.
[6] B. Lacoste, *Europe: Stepping Stones to Space*, Orbic, Bedfordshire, 1990; R. Jehn, *An Analytical Model to Predict the Particle Flux on Spacecraft in the Solar System*, Planetary and Space Science, Vol. 48, pages 1429–1435, 2000; Space Debris Subcommittee, *Position Paper on Orbital Debris, Edition 2001*, International Academy of Astronautics, Paris, 2001.

the high atmosphere is much greater near solar maximum conditions than at solar minimum.

Smaller pieces of debris are produced by the explosion, whether accidental or intentional, of upper stages or satellites. It has been computed that about half of the centimetre-sized debris has been produced in this way.[7] Military satellites are most responsible for this type of pollution, as stated in Chapter 1. Dangerous debris was produced when the core of the nuclear reactor of a military satellite was jettisoned in order to put it into a safe (higher) orbit at the end of the satellite's life. While doing this, the cooling system of the reactor let swarms of droplets of the coolant, a liquid sodium–potassium alloy, escape. These liquid droplets are dangerous sub-centimetric projectiles, which can penetrate the skins of satellites.

International treaties which forbid the intentional explosion of satellites are being prepared; they state that precautions must be taken against accidental events which may produce space debris. They also state that decommissioned satellites must be de-orbited and destroyed in the atmosphere or, if this is impossible as in the case of geosynchronous satellites, moved into a less-used orbit.

The most critical orbits are those lying between 1,000 and 1,400 km altitude, where the air drag is insufficient to cause debris to decay and re-enter. The ultimate danger is a situation in which there are so many objects that collisions are frequent enough to produce new fragments continuously. This sort of chain reaction would end up creating a debris belt in which no object could survive. But this nightmare scenario will not occur for several centuries.

Apart from the much-publicised accident when a *Progress* cargo craft hit a solar panel of the *Mir* space station, only one space collision between two unrelated objects has occurred to date. This was when a suitcase-sized fragment from the explosion of the upper stage of an Ariane rocket hit, after 10 years in space, the small French satellite *Cerise*. The damage in that case was not too large – a boom protruding

---

[7] L. Anselmo, B. Bertotti and P. Farinella, *Detriti spaziali*, CUEN Edizioni, Naples, 1999.

from the satellite for stabilization purposes was cut off – but the satellite might well have been wiped out.

The probability that a large piece of debris will come closer than 100 m from a satellite in LEO is about one in 100 years. In the case of a space station, a large piece of debris identified in advance can be avoided by suitable manoeuvres. Very small objects do not cause damage. Debris smaller than a few millimetres should be stopped by the shields of the *International Space Station*. The most dangerous particles are those which cannot be detected from the ground but are larger than a few millimetres. It has been computed that the probability that a 1 cm sized particle will pierce through the hull of the ISS space station is about 1% in its 20 years life, a risk that is worth taking. Tests in which astronauts have mended holes in a pressurized module have been conducted in large vacuum chambers, and have shown that they can manage the situation very well.

SPACE HABITATS

The *International Space Station* should have a crew of six, even if it may occasionally host a larger number. With time it could be enlarged, if the world's scientific and, above all, economic activities were to expand.

When the size of the space station is increased, the resident crew will also have to grow and an increasing number of people will begin to live permanently in space. Sometimes the space station and permanent bases on the Moon and on Mars have been compared with Antarctic scientific bases. In both cases they would be manned by scientists for stays limited in time, after which they would return to their usual occupations. However, important differences can be forecast. Antarctic bases generally work on a seasonal schedule, as during the austral winter they are either closed or their activity is reduced. Moreover, the investment costs are lower and a less intensive use is quite acceptable.

If and when industrial activities run from the space station increase, and if and when a large number of people live permanently in space, rapid turnover of staff would be both difficult and costly, and so very long stays will be encouraged. Perhaps the recent prolonged stays of some Russian cosmonauts on the *Mir* space station, in order to reduce launch costs, anticipate a future situation. Some people may start to consider a space station as their home.

The problems linked to weightlessness must not be understated and an artificial gravitational field should be created. As already said, the only way to do this is to rotate the whole space station to simulate weight with a centrifugal force. Neither *Mir* nor the *International Space Station* can do this and this limits the possibility of very prolonged stays. A future large space station will necessarily rotate about its axis.

A very large space station, with a complex ecosystem to maintain a breathable atmosphere, through the oxygen–carbon dioxide cycle due to plants, and to supply food, is usually referred to as a space habitat. The idea of building very large space habitats comes from the observation that an industrial civilisation makes a marked impact on the planetary ecosystem. The best way to solve the environmental problems of Earth could be to transfer some of the industrial activities and a large number of people into space. This idea was developed by Krafft Ehriche, whose concept of the space imperative includes a division of tasks between the terrestrial and space environments. More recently, one of the most active promoters of this idea was Gerard O'Neill, a physics professor at Princeton University, who designed space habitats able to host tens of thousands and even millions of people, giving them living conditions very close to those on the Earth.[8] Their construction will be possible only by exploiting resources from the Moon, and will require launch costs to be reduced by orders of magnitude. As a consequence, and contrary to what was believed by Ehriche and O'Neill, large space habitats are now regarded as a possibility only in the distant future. There is now a greater focus on the colonisation of Mars and the Moon.

[8] G.K. O'Neill, *The High Frontier: Human Colonies in Space*, Bantam Books, New York, 1977.

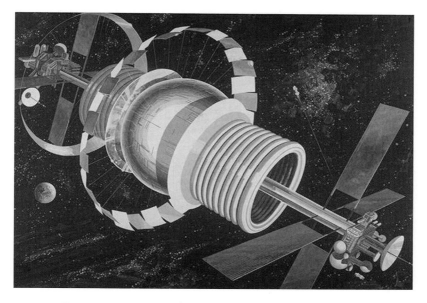

Figure 3.9. A conceptual space station where ten thousand people could live and work in space, similar to the *Island 1* project (NASA image).

Island 1 (Figure 3.9) is a project for a spherical habitat with a diameter of 460 m, able to host a population of 10,000, with a true ecosystem, with parks, small rivers, private houses, sports centres and all that can make life in space very similar to life on the Earth (Figure 3.10). *Island 1* could be positioned at either of the Lagrange points (L4 or L5[9]) on the orbit of the Moon, equidistant from the Moon and the Earth. Space habitats located so far from the surface of the Earth require good shielding from radiation, which could be made using material from the Moon.

The Lagrange points L1 and L2 of the Sun–Earth system have been proposed for large space stations. In this case they might host scientific communities, servicing a large astronomical and radio-astronomical observatory (at L1) or a solar observatory (at L2). A manned mission to one of these points has been proposed as the first 'deep space' mission with a human crew. It would prepare the ground

[9] See Appendix B.

Figure 3.10. Inside a space habitat (NASA image).

for these observatories, and test long-range human missions before setting sail for Mars.

Other projects considered by O'Neill deal with a cylindrical habitat, slowly rotating about its axis. One of these, 32 km long and with a diameter of 6.2 km, was designed for a population of 20 million (Figure 3.11).

Such habitats will tend to become economically self-supporting. They will certainly need to import raw materials from the Earth or, better, from the Moon, but they will export manufactured goods and perhaps energy and know-how. In a global economy, such as that already existing on the Earth, they will find their place and, if run profitably, they will flourish.

It is not unlikely that some will decide to reduce the artificial gravitation, either for economic reasons (as the structure has to be heavy to withstand the stresses due to rotation), or to simplify space operations in the vicinity of the space station and to get people used to a gravitational acceleration much lower than that experienced on

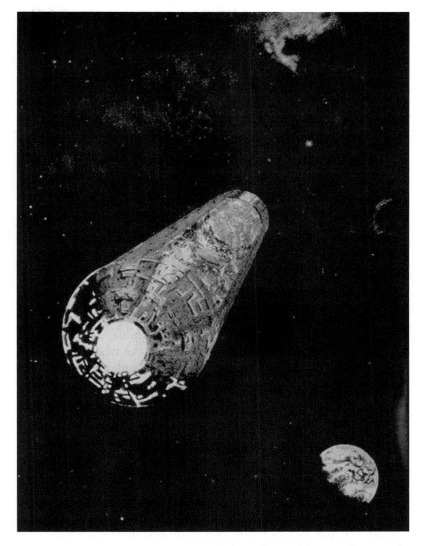

Figure 3.11. An enormous space habitat, designed for 20 million people, orbiting not far from the Earth (T.R. McDonough, *Space, The Next Twenty-five Years*, 1989. This material is used by permission of John Wiley & Sons, Inc.)

the Earth. The disadvantages of living in very low gravity conditions exist only for those who think in Earthly terms, but if people intend to spend their entire lives in a space habitat they can be better adapted to space conditions. The same also holds for ambient pressure. To reduce pressure while increasing the quantity of oxygen will perhaps be an inevitable choice to reduce structural stressing and the mass of the space station. If the partial pressure of oxygen is not much lower than that corresponding to sea level on the Earth, no adaptation will be needed. However, it is possible to reduce the atmospheric pressure, as is testified by people living in mountainous regions of the Earth who are perfectly adapted to their environment.

Long before these projects are implemented, a baby will be born on a space station. This may happen when problems preventing the journey back to Earth occur, or the birth is a premature birth, as has happened several times on board a ship or an aeroplane. And perhaps for the first time the baby will be carried back to the Earth as soon as possible, so that he or she can become used to 'normal' conditions, i.e. normal for those living on the surface of Earth. But those who grow up in reduced gravity conditions, and perhaps in a thinner atmosphere, will become different from people brought up in standard conditions on Earth. It is likely that they will be taller, with less-developed muscles and larger lungs. Other consequences for the human species of the colonisation of alien worlds are discussed in Chapter 10.

If a number of people live permanently in space habitats, there is no reason that those *cities in space* will be built only in near Earth orbits or in the Earth–Moon system. There are those who think that a habitat orbiting the Moon is a better solution for the exploitation of the resources of our satellite than a lunar base. This is, however, a controversial issue as the Moon has a gravitational field, even if weaker than that of the Earth. The Moon offers construction materials, perhaps water and above all protection from radiation and changes of temperature, using the dust abundantly present on its surface. All these materials could be launched into space from the Moon, to be used for space habitats, at a relatively low cost.

Figure 3.12. A space habitat moored to a small asteroid (T.R. McDonough, *Space, The Next Twenty-five Years*, 1989. This material is used by permission of John Wiley & Sons, Inc.)

When the asteroids start to be used as sources of raw materials, it could be more convenient to live in space habitats 'moored' to the asteroid than on its surface (Figure 3.12). The gravitational acceleration on the surface of an asteroid is so feeble as to be of very little advantage. It is easier to rotate an artificial habitat to simulate gravity than to rotate an asteroid. The velocity needed to leave the asteroid is so low that very little energy is required to move from the asteroid to the space station.

It is then possible that human expansion in space will occur mainly using artificial habitats scattered throughout the whole solar system. In such a scenario the human species could differentiate into 'Earthlings' and 'Space Children'; in the long run this could lead to two distinct species. As will be seen later, such a scenario is perhaps too limited with respect to possible long-term developments of human activities in space.

ENERGY GENERATION IN SPACE

One of the most critical aspects of all space missions is power generation. Power is required in space mostly for ancillary electrical equipment; in the future it may also be needed for the main propulsion system.

Solar cells have been used since the beginning of the space age; moreover, they are one of the products of space technology which have also found uses in more Earthly applications. A solar cell is a device which transforms the light from the Sun into electric power. As it has a rather low efficiency, only a small part of the energy from the Sun is converted, the remainder being reflected or transformed into heat, which must be dissipated by radiating it away to space. While the efficiency of the early cells was only a few per cent, now efficiencies higher than 10% are common and in the future this figure will be raised to 20%. The solar cells are connected to form panels, usually flat, extensible appendages of the spacecraft. Sometimes the cells are directly mounted onto the outer surface of the spacecraft.

The efficiency of the cells decays in time, due to spontaneous decay and to damage from radiation and the impact of micrometeoroids or space debris. A greater number of cells than that strictly needed must be available, so that enough power is still generated after some of them have been put out of use. Solar cells are often backed up by rechargeable batteries, which are charged when the power supplied by the panels is greater than that needed to supply energy for the equipment operating, for use when the spacecraft is in the shadow of some celestial body. This occurs particularly in the case of low Earth orbits, as the satellite spends somewhat less than half of the time in the Earth's shadow.

A solar cell generator for space use has a mass of about 1 kg for every 100 W of generated power and the hope is to halve that figure in the near future. The specific performance will thus be of about 200 W/kg. At the distance that the Earth is from the Sun (1 Astronomical Unit), the 'solar constant', i.e. the power of the electromagnetic radiation from the Sun, is almost 1.4 kW/m^2; with an efficiency slightly lower than 10% the power per unit surface of the panel perpendicular to the Sun's rays is 130 W/m^2.

Alternatively, to increase the efficiency of the system, mirrors could concentrate the light of the Sun onto a boiler, and then a turboalternator could convert the thermal energy into electrical energy. There is not, however, a distinct advantage in using such a complex system over panels of solar cells, at least for small- and medium-sized spacecraft. The optimum configuration depends upon the mass of the spacecraft and the power that has to be generated.

It is not always possible to use only solar cells. Fuel cells, for instance, which were developed for the *Apollo* missions, directly convert the chemical energy of a fuel and an oxidiser, such as hydrogen and oxygen, into electric energy. In fuel cells the fuel is not burnt and the energy is not transformed into thermal energy, as occurs in any thermal engine. The Space Shuttle uses fuel cells for its electric power generation. They are very well suited for short missions and their development for space use should also lead to their application on the Earth. Here the problems are mainly linked to their cost, but ground vehicles powered by electric motors and fuel cells have already been tested.

Neither power source is suitable for space probes which must operate for a long time at a large distance from the Sun. Fuel cells would require large quantities of fuel and oxidiser while the output of solar cells greatly reduces at increasing distances from the Sun. At the distance of Mars, their power is less than a half of that near Earth, and beyond the orbit of Jupiter their use is almost impossible.

In this case, and also for some other satellites, mainly military ones, radioisotope thermoelectric generators (RTGs) have been commonly used. They are small capsules containing a radioactive material such as plutonium 238, surrounded by a number of thermoelectric generators and then, on the outside, by a radiator. The radioisotope reaches a temperature which is higher than that of the radiator and this difference of temperature makes a current flow in the thermoelectric material. The efficiency of such devices is low, but they are compact, reliable and long lasting.

RTGs are thus a very convenient – and necessary – power source for space probes exploring the outer solar system, like the *Cassini* probe which is now heading for Saturn. The fears of using RTGs have been greatly exaggerated, as is demonstrated by the only accidents involving nuclear powered spacecraft to date.[10] The first occurred when the *Transit 5B-N3* satellite failed to attain its orbit in 1964. At that time the generators were designed to disintegrate in the high atmosphere, and the SNAP-9A RTG did. No measurable excess radioactivity was found. The second accident occurred when the launch of the *Nimbus B1* satellite failed in 1968. The SNAP 19 generator was this time designed to remain intact; it was recovered intact after 5 months in the ocean. And when the lunar module *Aquarius* of the ill-fated *Apollo 13* mission disintegrated in the Earth's atmosphere, its RTG went down intact into the ocean without any measurable radioactive contamination being found. The same happened when the launch of the *Mars 96* probe failed. The worst accident, and the only one in which there was contamination, occurred when the Russian *Cosmos 954* disintegrated in the atmosphere over an unpopulated region of northern Canada. Its large nuclear reactor (not an RTG, but a fully fledged reactor) disintegrated, and various radioactive fragments were found on the ground. The decontamination operation costing about US $8 million was paid for by the Soviet government. Subsequent *Cosmos* satellites had a provision for jettisoning the reactor, which was put in a safe higher orbit, before re-entry.

Similar to RTGs, but even less dangerous since they are far smaller, are Radioisotope Heat Generators (RHG). These are tiny radioisotope capsules which heat some crucial parts of a spacecraft which would otherwise have a too low temperature for correct operation. These are necessary for the thermal control of many spacecraft, particularly probes travelling far from the Sun.

The power available aboard a spacecraft has never been that high. The majority of space vehicles work with powers from a few tens

[10] For a detailed discussion of these fears, see Robert Zubrin, *Entering Space: Creating a Spacefaring Civilization*, Tarcher/Putnam, New York, 1999.

of watts to a few kW – from the power needed by a light bulb to that needed by a washing machine. The radio transmitters of the *Pioneer* or *Voyager* probes which are now outside the solar system have a power of a few watts: it is almost incredible that it is possible to receive messages broadcast with so little power from those astronomical distances.

Large space systems operate with remarkably low power. The eight enormous solar panels of the *International Space Station*, for instance, will generate a power of 75 kW. The space station has to operate on a total power which is equivalent to that of the engine of a medium-sized car.

To understand why it is so difficult to generate high power in space it is enough to remember that the majority of electric power stations on Earth transform the energy of a primary source (chemical energy of a fuel, or nuclear energy) into heat, which is then converted to mechanical energy through a thermal engine (steam or gas turbine, reciprocating engine) and then into electric energy through an electrical machine (usually an alternator). The most critical phase is the thermal energy–mechanical energy conversion whose efficiency is not great and is limited by a theoretical maximum, the efficiency of the Carnot cycle. The latter increases with the ratio between the maximum temperature of the working fluid (steam, or gases obtained from combustion) and the temperature which it has after the energy conversion.

Since the maximum temperature is limited by the characteristics of the materials which are in contact with the working fluid, to increase the efficiency it is necessary to reduce the temperature at the outlet of the thermal engine. Power stations are therefore located near rivers or lakes, which can supply large quantities of cooling water, or have cooling towers, in which smaller quantities of water cool the working fluid by evaporation. But in space there is no cooling fluid and it is impossible to throw away water or other fluids to dissipate heat. The only way is to radiate away the waste heat, and this requires either large radiators or high outlet temperatures.

If the temperature is increased, the efficiency drops and large quantities of thermal energy must be produced to use just a small fraction of it, while if the temperature is decreased the radiator becomes very large and heavy. A trade-off study must be performed and an optimum found. In any event, every device which generates power in space using thermal energy is heavy, has a large and rather delicate radiator, and has an efficiency which is lower than that of an equivalent device on Earth.

Different solutions to this problem have been proposed, but at present little can be done practically to improve matters. An increase of the high-temperature strength of some materials would allow the temperature of the working fluid to be increased, improving the efficiency. Direct energy conversion devices, which are not based on rotating machinery, such as MHD (magnetohydrodynamic) devices, would similarly increase the efficiency. The best solution would be the direct conversion from chemical to electric energy, as in fuel cells, or from nuclear to electric energy. All these measures, which would also be useful for Earth-based energy systems, could improve substantially the energy available in space.

All devices based on the chemical energy of a fuel are limited to short missions, as in the case of the *Apollo* spacecraft or the Space Shuttle, as both fuel and oxidiser must be carried into space. The only energy source suitable for long missions is nuclear energy. The use of a nuclear reactor in space is surely even less questionable than on Earth, as in space the radiation level is already high and the presence of a reactor would not make things worse; moreover, radioactive waste can be disposed of in simple and safe ways. As already stated, the only risks are linked with accidents in the launch phase, but the small quantity of fissionable material needed can be packed in an accident-proof container. Accidental re-entry of a satellite with a nuclear reactor on board can cause radioactive contamination, but this can — and must – be avoided. For this reason nuclear reactors are more suitable for spacecraft leaving the Earth than for Earth-orbiting satellites. The decrease of the funding for research in nuclear energy utilisation caused delays in this sector: only 20 years ago it was taken for granted that by the end of the twentieth century large space stations powered by nuclear reactors could be built. The United States built several reactors of the SNAP class, but at present the largest space nuclear reactors are the Russian *Topaz*.

Most experts, even those who oppose the widespread use of nuclear energy in space, think that it is not advisable to seek an agreement to ban nuclear reactors and RTGs from spacecraft. This would slow down space research and exploration too much, putting an end to absolutely safe missions such as sending probes to the outer solar system. Besides, it is absolutely not necessary.

A very welcome alternative would be the use of controlled nuclear fusion, but the large research effort in this field has not yet led to practical applications. It is hard to predict with confidence when these might be achieved.

However, the problems linked with power generation are more critical for the construction of electric-propulsion spaceships, as will be shown in the following chapters, than for space stations. Spaceships must be accelerated to very high speeds and their mass must be as low as possible. While space stations must remain in orbit an increase of their mass increases the launch costs but does not greatly affect their performance. At present, the simplest way to increase the amount of power available on space stations seems to be to install larger solar panels. The cost per unit area of solar panels and their mass/power ratio has constantly decreased, and their efficiency increased. There are no theoretical limits to the use of large or very large solar panels in Earth orbit. The structure needed to support solar panels in orbital applications is minimal, even though it may be difficult to keep very large panels oriented to face the Sun. Also the force exerted on them by the sunlight has to be reckoned with.

ORBITAL POWER STATIONS

The high efficiency of solar cells in space and the possibility of building very large arrays (even of several square kilometres) of panels in orbit led to the belief that it would be possible not only to satisfy all the needs of space stations using energy from the Sun, but also to produce

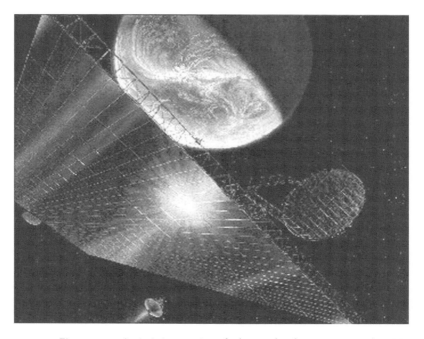

Figure 3.13. Artist's impression of a large orbital power station (NASA image).

surplus energy which could be sent down to the Earth (Figure 3.13). This could be very attractive, and is technologically feasible. The energy obtained from large solar panel arrays or, as an alternative, from a solar thermal generator, can be converted into a microwave beam, sent to the Earth and then transformed back into electric energy. The conversion efficiency can be quite high and, if the efficiency of the cells themselves is not accounted for, greater than that of a power station on the ground. In space there is plenty of energy from the Sun and the process is intrinsically very clean. Energy generation in space (SPS, Space Power Systems) or on the lunar surface (LPS) and its transmission to the Earth is one of the basic – and attractive – features of the space option.

The generation of energy in space to satisfy, at least partially, the energy needs of our planet could substantially reduce our use of non-renewable energy sources and reduce pollution. Moreover, generating energy in space avoids covering large areas of land with solar panels. Further, large areas of solar panels on the surface of the Earth would cause significant thermal pollution, as their low efficiency would cause the absorption of huge quantities of heat from the Sun which has to be dissipated locally. On the contrary a space system produces only a reduced thermal pollution, as the energy which the panels do not convert into electric energy is directly radiated back to space. The microwave beams can be converted back into electric energy with very high efficiency and little heat is produced in the process.

An orbital solar power station in geostationary orbit would require far smaller panels than an equivalent system on the Earth, because of the larger efficiency of the solar cells in space and for the obvious reason that a solar power station on the ground can work for only about 12 hours each day, while in space it can generate electricity for 24 hours. The danger that the microwave beam falls outside the receiver antenna and causes damage is not great, both because it is easy to take the correct measures to avoid this risk and because the power density of the beam is not very high; it would not cause severe damage even in the worst of circumstances.

If this energy source is so clean and almost unlimited, why is it not yet used on a large scale? One of the reasons often put forward by the supporters of the space option is the habit of searching for the solution to the problems of our planet on the Earth itself, neglecting the immense potential of extraterrestrial resources. But there is also a technical and an economic reason. A large quantity of materials and devices must be taken into orbit to build a power station and the many tasks of assembly, monitoring and maintenance must be performed there. The cost of such an operation is very large. The availability of cheaper launch vehicles and the experience obtained by operating space stations are prerequisites for the construction of these large scale space facilities.

The most convenient way to build the large solar panels of the power stations is to use materials mined on the Moon: lunar regolith is very rich in silicon, the basic constituent of the solar cells, containing about 20% silicon. The low escape velocity from the surface of the Moon means that this construction material would be put into Earth orbit at a cost which is just a fraction of that involved with launches from our planet, particularly if electromagnetic launchers are used. To implement the space option, at least for the part linked with space power stations, humankind must first go back to the Moon, build permanent bases on its surface and start its exploitation.

Small-scale experiments of power generation in space and of its transmission to the Earth would be a welcome step in the right direction.

LIGHT FROM SPACE

Large reflecting structures could be located in Earth orbit to reflect light from the Sun onto the surface of our planet. This has been suggested several times, particularly by the Russians, who see this as a way of illuminating towns and high latitude regions at night time and of saving energy.

Mirrors can be built using very lightweight material, like thin plastic film covered by a few microns of aluminium. The film must be kept tight by a structure which might be an inflatable one or might be based on the deployable beams which have already been tested in space. An alternative would be to spin the whole system so that the mirror has the correct shape. Mirrors of this type are not dissimilar from the solar sails which are being tested as propulsion devices (see Chapter 4), but the low mass requirements are much less severe in the case of orbiting mirrors than for sails.

The easiest solution to illuminate a certain part of the Earth is to locate the mirror in a geostationary orbit. However, the distance from the Earth and the difficulty in controlling the shape of the mirror within strict tolerances lead to a large illuminated zone. For instance, a 340 m diameter circular mirror could illuminate an area of 340 km diameter with an illumination level equal to that of the full Moon. A preliminary study performed by Russians and funded by the

European Community (*Light from Space* project) suggests the use of a satellite with an array of ten such mirrors to supply night-time illumination over a zone of 340 km diameter. The total power reflected back towards the Earth is about 1 GW, the output of a large electrical power station.

Another suggestion is to use mirrors in low Earth orbit. This involves having several satellites and manoeuvring them in such a way as to illuminate the required area. With such a system the light intensity is much larger, for a mirror of the same size, and the zone on which the light is received is smaller and more controllable.

Many oppose such projects for their environmental impact. At first sight it might seem that a number of such mirrors would contribute to global warming, since they reflect onto the Earth energy from the Sun which would otherwise be lost to space. This is, however, not correct if the energy needed to illuminate the same area had to be generated using non-renewable sources. Owing to the low efficiency of the conversion from thermal energy to electric energy and then to light, to obtain the same effect a much larger quantity of energy, and hence of heat, would need to be produced on the surface of the Earth, not to mention the other types of pollution (chemical, greenhouse-effect gases, and so on, depending on the primary energy source). Reflecting energy from space, from this viewpoint, would be an environmentally friendly practice.

Nevertheless even the simplest 'artificial Moon' is not free from drawbacks. The presence of a new source of light in the sky makes astronomical observations much more difficult. The light pollution problems which plague astronomy would be made much worse.[11]

---

[11] D. McNally (editor), *The Vanishing Universe. Adverse Environmental Impacts on Astronomy*, Cambridge University Press, Cambridge, 1994.

4   Robots in the solar system

Robotic spacecraft have visited every planet in our solar system except Pluto. They have sent back a large number of fascinating images as coded radio signals. They have viewed – close up – many planetary satellites and the ring systems of the four giant gas planets. In the relatively near future new propulsion systems could take a variety of novel equipment to interesting places such as Mars, Jupiter's enigmatic moon Europa, comets or asteroids, or the remoter regions of the solar system, the Kuiper belt or even to the Oort cloud.

LARGE INTERPLANETARY SPACECRAFT

Since the 1970s, large spacecraft have visited all the planets of our solar system, except Pluto, and many of their satellites. The observations which they made have advanced planetary astronomy beyond all expectations. Robotic spacecraft had earlier studied the lunar surface in sufficient detail to prepare for the landing of the *Apollo* astronauts (Figure 4.1).

Two very important robotic missions were the *Mariner* and *Magellan* probes which, for the first time, studied the surfaces of Mercury and Venus. These cannot be observed by telescopes either on Earth or in Earth orbit. The *Voyager 1* and *Voyager 2* probes, after crossing the whole outer solar system, studying Jupiter, Saturn, Uranus and Neptune, left the solar system and are now travelling towards interstellar space. Also, *Pioneer 10* and *Pioneer 11* have already crossed the orbit of Neptune, leaving the solar system (Figure 4.2). Since the orbit of Pluto is highly elliptical and crosses the orbit of

Figure 4.1. The *Surveyor III* probe visited by *Apollo 12* astronauts on November 20, 1969. The Lunar Excursion Module is seen in the background (NASA photo).

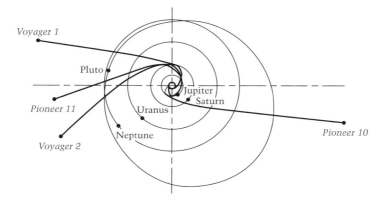

Figure 4.2. Projection of the trajectories of *Voyager 1*, *Voyager 2*, *Pioneer 10* and *Pioneer 11* onto the ecliptic plane. The positions of the probes and the planets are shown for the year 2000. (E.F. Mallove & G.L. Matloff *Starflight Handbook*, 1989. This material is used by permission of John Wiley & Sons, Inc.)

Figure 4.3. Artist's impression of the *Galileo* probe (NASA image).

Neptune and, at present, Pluto is nearer the Sun than Neptune, all four space probes may be said to have left our solar system. They are heading for the heliopause where the interstellar medium meets the interplanetary medium, the solar wind flowing away from the Sun.

It may seem strange to non-physicists to distinguish between the interstellar medium and the tenuous matter which is encountered in the outer reaches of the solar system. After all, both are practically a vacuum, and a better vacuum than the best obtainable in laboratories on Earth. But the heliopause is an interesting boundary between rarefied matter of different origins.

Voyager, *Mars Observer* (which unfortunately was lost at the end of its journey to Mars), and *Galileo* (Figures 4.3 and 4.4), which after an eventful journey made fascinating observations of Jupiter's satellites, were very large probes, extremely complicated and constructed at a very high cost. A probe of this type has a mass of 3 or 4 tonnes, and a cost of US $ several billion, launch included. The last probe of the large type is *Cassini*, which is bound for Saturn, carrying on board the landing module *Huygens* (Figure 4.5). This will enter the

Figure 4.4. A painting of the *Galileo* entry module approaching Jupiter (NASA image).

Figure 4.5. The *Huygens* lander on the surface of Titan, a moon of Saturn (painting by Michael Böhme).

atmosphere of Titan, the largest of Saturn satellites, and land on its surface in 2004.

All these large space probes are American, except the landing module *Huygens* which has been built by ESA. Initially, Europe was little involved in the automatic exploration of the solar system but, from the mid 1980s, its activity in this field increased with the launch of *Giotto* and other probes. *Giotto* paid a very successful visit to Halley's comet in 1986 and, after hibernating for more than 2 years, was woken up and directed toward the comet Grigg–Skjellerup. This second successful mission had never been contemplated in the design stage, and the fact that a space probe could be used again after completing its useful design life is remarkable.

The Russians had earlier started a very important series of launches, towards the Moon and then to Venus. Their activity then slowed down; the failure of the *Mars 96* probe to head for Mars from its orbit 140 km above the Earth jeopardised a programme which was already in a difficult situation for lack of funding.

LOW-COST SPACE PROBES

In the 1990s, many things started to change. The *Mars Observer* failure, the dangers which jeopardised the *Galileo* mission and, above all, reactions to the bureaucratisation of various space agencies gave strength to those advocating a new approach to space. This led to the call for 'faster, better, cheaper' missions based on the concept of reducing the cost of a mission by shortening the development time and by simplifying the various subsystems of the spacecraft. Underlying this approach is the consideration that bureaucratic procedures unnecessarily stretch out programme times and cause an increase in costs, neither of which is justified in technological terms. Cheaper and faster may thus be joined with better. All this is possible at the cost of reducing the reliability of the probe which, in the case of automatic missions, may not be a real consideration. If the mission cost is halved, it would be possible to launch two spacecraft of the same type for the same mission, and so increase the chances that it is carried out successfully. In this way the chances are reduced that an unlikely – but unavoidable – event, such as collision with a micrometeoroid or piece of space debris, wipes out an otherwise very reliable probe and terminates the mission.

The 'faster, better, cheaper' philosophy is also based on the rapid advances made in electronics and computer science, especially in the miniaturisation of subsystems and in more efficient control and information processing algorithms.

The first offspring of this new philosophy was the *Clementine* probe, built by the US Air Force and launched from Vandenberg base in California with a non-reusable rocket. Just 2 years and US $80 million were needed to build it, compared with 4 to 10 years and US $2 to 3 billion for the large NASA probes of the 1960s and 1970s. The mission included a prolonged stay in orbit around the Moon and then, after leaving the orbit, a journey to meet the asteroid Geographus. The first part of the mission was a success beyond all expectations – almost one million very detailed pictures, with a resolution of about 200 m, were taken of the lunar surface. Maps of the Moon's surface now exist which are even more detailed than those made by the astronauts of the *Apollo* missions. While over the polar regions of the Moon, which are still practically unknown, *Clementine* found clues which might indicate the presence of water (as ice) there. This result is very promising for the construction of future permanent bases on the Moon.

Unfortunately, after leaving its orbit around the Moon, the probe stopped working; just why is not clear. In this mission we can see the advantages and also the drawbacks of the faster, better, cheaper, approach; important results were obtained at a low cost, but also a reduced reliability. It is, however, true that if two *Clementine* spacecraft had been launched, the cost would still have been a fraction of the cost of a large probe, but the chances of completing the whole mission would have been far greater.

Clementine also demonstrated that it is essential to incorporate the most advanced technologies in space probes, and to depart from that conservative approach which is typical of large agencies.

The sensors and many subsystems were directly derived from those developed in the SDI (*Star Wars*) project. Also, the materials used in much of the structure, the guidance systems and the software were all innovative. A new Moon probe which will follow the same approach as *Clementine* is *Blue Moon*, built by the United States Air Force Academy, with the co-operation of students, and due to be launched soon.

Clementine and *Blue Moon* are perhaps the most extreme cases, but other recent and future NASA robotic missions follow a similar approach. The success of the *Pathfinder–Sojourner* probe, whose lander and rover performed outstandingly on the Martian surface, is an important point scored by the supporters of the new philosophy. The figures are impressive: the *Pathfinder* mission cost US $250 million and took 4 years, while the *Viking* missions cost US $3 billion (in 1997 dollars) and required 8 years work.

It must, however, be remembered that low-cost probes can yield only a fraction of the scientific results of the large probes for which they are substitutes, even if miniaturisation can partially reduce this drawback. Those who oppose the faster, better, cheaper approach hold that it is deceptive and even dangerous to try to save money in this way, as the disappointment due to the smaller number of scientific results could jeopardise the very concept of robotic space exploration.

The only way to avoid this risk is to increase the number of missions, in such a way that their quantity can outweigh the reduced importance of each. What is needed is a programme with a good number of specialised missions, each one focused on a few well-identified objectives, but well coordinated with each other. But perhaps the real point of the faster, better, cheaper approach is that it contemplates a larger probability of failure, which is more than compensated for by the reduction in cost. However, when a failure actually occurs, as in the case of *Mars Polar Lander* and *Mars Climate Orbiter*, all that the public realises is that the mission has failed and that taxpayers' money has been wasted. This leads to a decrease of confidence not only in the space agency but also in technology and science in general, both of which may lead to a reduction of future funding. The fact that the reason for failure seems to be 'stupid' (could there be an 'intelligent' failure?) makes things worse. Failure in the faster, better, cheaper approach is likely to be due to human error. *Mars Climate Orbiter* was doomed because of a failure in communication between two engineering teams using different units of measurement. In the case of the *Mars Polar Lander*, radio communications were lost near Mars. Such failures can be ascribed to cutting funds (cheaper) and the time (faster) needed for cross-checking and testing everything, thoroughly.

At any rate this is NASA's new approach to Mars exploration. ESA seems to be more reluctant to apply the new approach, partly because probes like *Giotto* were smaller and cheaper than the American probes anyway. The European *Mars Express* mission is, however, the cheapest Mars mission ever designed.

PROPULSION IN DEEP SPACE

The quantity of propellant used to obtain a given thrust for a given time is reduced by increasing the velocity of the jet.[1] As the velocity which can be obtained using chemical reactions is limited, chemical rockets require very large quantities of propellant when used for missions involving large velocity increments, termed Δv. To reduce the flight times for deep space without requiring such large quantities of propellant, propulsion systems based on a different principle must be developed. In particular, the two functions of power and thrust generation must be separated.

An alternative to a chemical rocket is a nuclear fission rocket. A nuclear reactor transforms the nuclear energy of a fissionable element, for example, uranium 235, into thermal energy, which in turn heats a fluid which is expanded in a nozzle, generating thrust. Higher ejection velocities can be obtained in this way, even if at the cost of greater technological complexity. The nuclear reactor must work at temperatures which are higher than those of reactors used in typical power stations, and even greater than the melting point of uranium (1,133 °C).

---

[1] See Appendix C.

Despite the difficulties, both the Americans and the Russians were close to developing a viable nuclear rocket at the beginning of the 1970s. The American NERVA (Nuclear Engine for Rocket Vehicle Applications) project was abandoned for economic and political reasons, when the nuclear rocket engine Kiwi had already been ground-tested with success.

The combination of high exhaust velocity and high thrust makes nuclear rockets unique among deep-space propulsion systems, and they should prove to be an enabling technology for future manned missions in the solar system. Even if it is possible to design a mission to put astronauts on Mars using chemical propulsion, a nuclear engine would make the journey faster, reducing the dangers for the crew and increasing the payload taken to the planet's surface, and so enhancing the scientific return of the mission. The Mars Reference Mission outlined by NASA in 1997 (see Chapter 6) is based on the use of a nuclear thruster, derived from the old NERVA rocket, for the injection into the trans-Mars trajectory (Figure 4.6).

To increase the temperature, and hence the velocity, of exhaust gases, gas-core reactors have been suggested but, after the initial studies performed in the 1960s, they have been practically discontinued. The high thrust of nuclear rockets makes them the only form of high exhaust velocity device able to lift off from a planet. Safety reasons and the difficulty of avoiding radioactive exhausts will probably prevent their being used in the Earth's atmosphere, at least for the first stage; their use in deep space does not have serious drawbacks, however.

Another type of propulsion is thermal solar propulsion: light from the Sun is concentrated by mirrors or lenses and heats a fluid, which produces a high velocity jet. Systems of this type have been suggested for trajectory correction or to transfer a satellite from low Earth orbit to geosynchronous orbit. There are projects planned to experiment with solar thermal thrusters in space in the coming years.

Another possibility is the electric thruster. An electric current heats a fluid which, having been expanded in a nozzle, constitutes the jet. Electrothermal propulsion can be implemented by using a

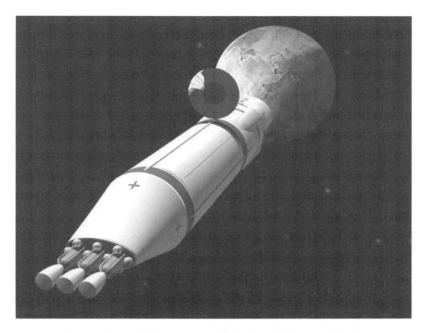

Figure 4.6. Artist's impression of the Mars Reference Mission spacecraft approaching Mars. The nuclear thrusters are clearly visible (NASA image).

resistor (resistojet) or, better, an electric arc (arcjet). To increase the ejection velocity further a different principle must be used, such as direct acceleration by means of electric or magnetic fields as in plasma and ion engines.

A further type of electric propulsion is based on electromagnetic launchers to accelerate pieces of solid material and to exploit the reaction on the thruster to propel it.[2] The idea is most applicable when the mass to be ejected is very cheap, as in the case of lunar material launched into space by an electromagnetic launcher, and when electric energy is abundant at low cost. This could therefore be the most suitable propulsion system for service vehicles near space habitats and for spaceships moving between these habitats and

[2] G.K. O'Neill, *The High Frontier: Human Colonies in Space*, Bantam Books, New York, 1977.

orbits around the Moon, although it has the drawback of producing a considerable quantity of space debris.

Ejection velocity is, however, not the only important feature of a propulsion unit. The power needed to accelerate the fluid, and consequently the mass of the power generator, increases with increasing jet velocity. Because a decrease of propellant mass is accompanied by an increase of the mass of the propulsion system, it is possible to find a value of the ejection velocity which causes the total mass of the vehicle to be a minimum, for a given mission.

The fields of optimum application of chemical and electric propulsion are different. In the first case the mass at liftoff is mostly the mass of the propellant while the engine mass is small; the propellant is burnt and ejected in a relatively short time (a few minutes), producing very large thrusts. Such systems are often said to be energy limited, as their greatest limitation is the total quantity of energy stored on board in the form of chemical energy.

In the case of electric propulsion, and particularly for ion engines which can achieve very high ejection velocities, the limitation comes mostly from the mass of the electric generator, which is proportional to the power needed to produce the jet. These systems are said to be power limited, as usually the total energy available is not a major problem. If solar cells produce the electric energy, the energy available on board is practically unlimited as it comes from the Sun; the limit coming from the propellant stored on board is not very severe, and the mass of fluid ejected can be quite small as the ejection velocity is very high. The available power produced by the solar cells, on the contrary, constitutes a severe limit to the total amount of thrust which can be generated. Electric thrusters are thus suitable for producing low thrusts, sometimes very low thrusts, but for long times. For these reasons, electric propulsion cannot be used to leave the surface of the Earth: it is practically impossible to produce a thrust which matches the weight of the spacecraft at lift off.

Small spacecraft operating in the inner solar system can, with advantage, use solar-electric propulsion. Solar panels supply the power to energise the electric, possibly ion, thrusters. For large spacecraft, nuclear-electric propulsion seems to be preferable; a nuclear reactor generates the electric power needed to operate ion or plasma thrusters, the latter being very well suited to large vehicles. The difficulties of building light and reliable space power generators have limited this approach. Thus research efforts in this direction must be intensified if the goal of achieving some mobility in the solar system is to be reached. Of course, solar-electric propulsion becomes less and less effective with increasing distance from the Sun, so that nuclear-electric propulsion will be the only way of travelling through the outer solar system.

Controlled nuclear fusion could be another enabling technology, both for power generation for electric propulsion and for the direct production of a jet of extremely high velocity. Conceptually, such thrusters are not very different from the fission rockets already mentioned, but for the fact that a small quantity of deuterium or helium 3, confined by magnetic fields at the centre of the fusion chamber, is substituted for the core of fissile material such as uranium 235 or plutonium. The energy produced by nuclear fusion heats flowing hydrogen gas, which is expanded in a nozzle and constitutes the jet. The products of the nuclear fusion may also be ejected to contribute to the thrust.

Ejection velocities far greater than those typical of fission rockets can be theoretically obtained, but there is another important advantage. Using helium 3, the jet contains very few neutrons and radioactive elements. It is essentially a 'clean' nuclear engine. Nobody has yet succeeded in producing significant power from controlled thermonuclear fusion. And it is uncertain whether, if and when that is done, it will be possible to build a system light enough to be used as a space thruster. But it may even be easier to control thermonuclear fusion in a space thruster than in a reactor suitable for energy production on the Earth's surface.

Pulsed nuclear propulsion can, on the contrary, be achieved using present technologies. In its simplest form it can be implemented by exploding several small nuclear charges behind the space vehicle; the products of the explosion would hit a plate placed at its tail, pushing it forward. A number of such explosions would allow very high speeds to be reached. The Orion project, based on this principle, was started in 1958 and the design group, led by Freeman Dyson, performed a very detailed study. When the project was interrupted in 1965, a final design had been agreed and a small rocket, in which the nuclear explosions had been substituted by small charges of a chemical explosive, had been flight tested. The experimental rocket, named Putt-putt, was launched from the ground and reached an altitude of 60 m, about the same altitude as was reached by the first liquid propellant rockets built by Robert Goddard. It is worth noting that the test was performed on the ground, in a strong gravitational field: no other propulsion device with high ejection velocity could work under these conditions.

The Orion spacecraft would have used small nuclear fission charges, each with a power ranging from 0.01 to 10 kilotonne TNT equivalent, loaded with inert material which, hitting the back plate attached to the vehicle through shock absorbers, would produce the forward thrust (Figure 4.7). The Orion project raised much enthusiasm. The proposers suggested that this method could completely avoid the stage of chemical propulsion spaceflight, switching to nuclear interplanetary flight. Their plan was to launch the first interplanetary nuclear spaceship in 1968.

Although the project was abandoned, the idea of nuclear pulsed propulsion was further developed, mainly with nuclear fusion in mind. In this case there is no need to wait for controlled thermonuclear fusion, as very small fusion bombs already developed by the military could be used. Soon the advisability of having a large number of very small bursts became clear. The concept of microexplosions, each with a power of a few tonnes of TNT equivalent, would simplify the structure and reduce both the mass and the cost of the spacecraft. This propulsion scheme would be suitable for the Daedalus project (described in Chapter 8) for a mission in interstellar space;

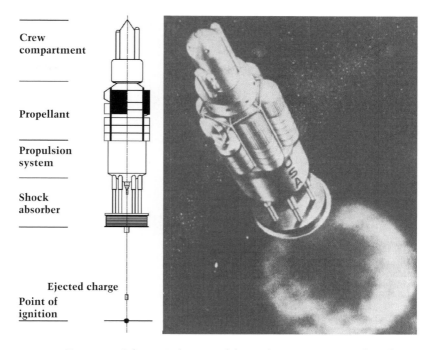

Crew
compartment

Propellant

Propulsion
system

Shock
absorber

Ejected charge
Point of
ignition

Figure 4.7. Schematic diagram of the nuclear Orion spacecraft, and an
artist's impression of it (NASA image).

pulsed nuclear propulsion devices would accelerate a spacecraft to
high enough speeds to leave the solar system.

The concept has been proposed again with the ICAN II[3] system.
Here, very high temperatures are generated by the annihilation of a
few antiprotons hitting a natural uranium target to start the fusion
of hydrogen microspheres. The principle seems to be very promising,
as the quantity of antimatter needed is very small, orders of magnitude smaller than that needed for a true antimatter propulsion device (see Chapter 8). A spaceship for interplanetary missions which
would travel between the Earth and Mars in 45 days with a consumption of just 100 ng (0.0001 milligrams) of antimatter, has been studied

[3] R.A. Lewis *et al.*, *Antiproton Catalyzed Microfission/Fusion Propulsion Systems for
Exploration of the Outer Solar System and Beyond*, First IAA Symposium on
Realistic Near-term Advanced Space Missions, Torino, June 1996.

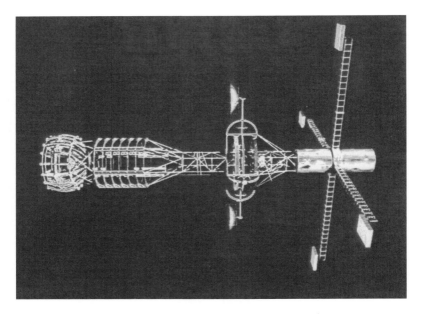

Figure 4.8. The proposed nuclear propulsion spacecraft ICAN-II.

(Figure 4.8). This quantity is of the same order of magnitude as that produced by the Fermilab in one year. The system seems to be feasible with present technologies and it could be extended to interstellar propulsion, provided that the ejection velocity can be increased.

At the end of the Cold War, the idea of using huge thermonuclear explosions for space propulsion was again put forward. The nuclear weapons of the superpowers would have to be destroyed, following disarmament agreements, and perhaps that could be done in space to accelerate a spacecraft to a large velocity. Whilst this idea is very appealing, it is not really practical.

The energy from the Sun can also be used to travel in the solar system in another way. Light from the Sun exerts a pressure (so-called radiation pressure) on all bodies in space. This effect is usually considered as one of the causes which perturb the trajectory of a spacecraft, particularly those having a small mass and a large surface area. Using a large reflecting surface, a 'solar sail', radiation pressure due to the sunlight can propel a spacecraft. But this radiation pressure due to the Sun is very weak: at the distance of the Earth from the Sun it is slightly less than 10^{-5} N/m^2, i.e. one tenth of a billionth of the atmospheric pressure at the surface of the Earth. This means that a square solar sail of 100 m side collects a thrust of only 0.1 N (10 grams). Although this thrust is very weak, it is available free, and can be applied for very long times. Moreover, by changing the orientation of the sail, the direction of the thrust can be controlled. Though the light travels radially away from the Sun, a solar sail can also move a spacecraft inwards, in a way which is similar to sailing on the sea against the wind.

The sail must be very light in weight, a very thin aluminium foil, for instance, thinner than the already very thin aluminium foils used for keeping foods fresh. The sail must be kept taut using masts and booms, themselves very light, and kept in place by thin stays. Different geometries, such as square (Figure 4.9), circular and lobed sails, have been studied. The possibility of rotating the whole structure to use centrifugal force to achieve the required stretching has even been investigated[4].

Solar sails have been considered for a number of missions, such as a probe to Halley's comet. But the only actual space demonstration was that performed from the *Mir* space station (Figure 4.10): in February 1993 the circular *Znamia-2* sail was deployed from an automatic *Progress* cargo craft. Different societies – the American World Space Foundation, the French U3P (Union Pour la Propulsion Photonique) and the Japanese Sail Consortium – actively promote this method of space propulsion.

At the beginning of the 1990s a regatta for sailing space probes on the Earth–Moon route was proposed to celebrate the 500th anniversary of the landing of Columbus in America (the Columbus 500 Cup). Three probes, representing Europe, the Americas and Asia, were to participate. Unfortunately, the idea could not be put into practice for

---

[4] J.L. Wright, *Space Sailing*, Gordon and Breach Science Publishers, New York, 1994; C.R. McInnes, *Solar Sailing: Technology, Dynamics and Mission Applications*, Springer-Praxis, Chichester, 1999.

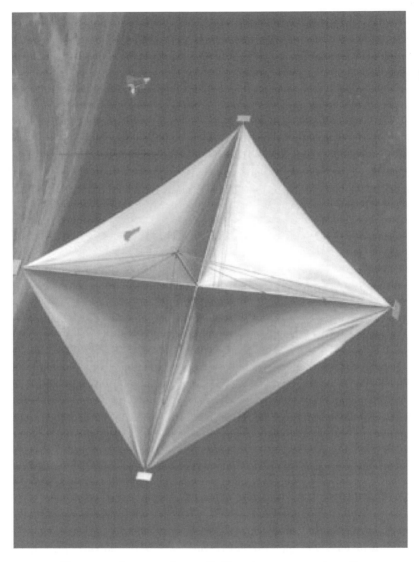

Figure 4.9. A spacecraft propelled by a large square solar sail
(J.L. Wright, *Space Sailing*, Gordon and Breach Science Publishers,
New York, 1994).

Figure 4.10. The solar sail *Znamia-2*, with a diameter of 20 m, unfurled from an automatic cargo vehicle *Progress* near the *Mir* space station. The sail, made of eight triangular panels, was kept taut by rotation (Space Regatta Consortium, Energia, image).

lack of funding. The idea was later revived and again abandoned, and the same fate befell a proposal put forward within ESA for a mission, based on a solar sail, to celebrate the new millennium.

A privately funded project is the *Cosmos 1* solar sail, developed by the Planetary Society with the OneCosmos Network, a media and Internet company. The aim of the project is to demonstrate in space the deployment of a small solar sail, firstly during a suborbital flight and then in orbit. In the orbital flight, an increase of the orbit due to the pressure of sunlight should be measurable. The first suborbital test was performed in July 2001, using a Volna rocket, a converted ballistic missile launched from a submarine. It failed due to the unsuccessful separation of the sailcraft from the launcher. Other tests are planned soon.

It is also possible to use the solar wind, the stream of charged particles continuously emitted by the Sun, with a very large loop of superconducting material in which a large electric current circulates, i.e. a magnetic solar sail or *magsail*, to generate the thrust. If the current is high enough, a thrust which is larger (for an equal sail area) than that due to the sunlight may be collected; this thrust decreases more slowly than that due to the radiation pressure with increasing distance from the Sun. But magnetic sails have a basic drawback – the strength of the solar wind is extremely variable in time. Without knowing the available thrust in advance it is impossible to design a space probe trajectory. For this reason, and for the greater complexity of the device, magnetic sails are still just a hypothetical alternative to more conventional propulsion techniques.

Recently, the idea of exploiting the solar wind has been revived by a study on magnetic bubbles, funded within the NASA Advanced Transportation Plan. A magnetic field generated by the spacecraft, with the help of a small quantity of ionised gas (plasma) ejected from it, creates a sort of small magnetosphere around the vehicle, qualitatively similar to (although much smaller than) that surrounding a planet with a strong magnetic field, like the Earth or Jupiter. The solar wind interacts with this magnetic bubble and, being deflected by it, produces a thrust. The bubble produced by a 1 kW magnetic field generator would, at the distance of the Earth from the Sun (1 AU), be of 15 km diameter. Less than 1 kg of helium would be consumed per day to produce the plasma. Such a bubble would produce a thrust of 1 N or so, quite a large thrust for a device using the pressure of the solar wind.

Some experiments conducted in large vacuum chambers at NASA's Marshall Space Flight Centre have shown that the system actually works, even though the limited size of the chamber meant that only a small bubble was produced. This Mini-Magnetospheric Plasma

Propulsion (M2P2) method has an added advantage. Just as the Earth's magnetosphere protects us from energetic charged-particle radiation from the Sun and outer space, the spacecraft is protected from such radiation. Another advantage is that it seems to provide a constant thrust: for a less powerful solar wind, the more the bubble expands (at least within certain limits). As a result, the thrust which the vehicle receives does not change much with distance from the Sun or with the intensity of the solar wind. If future tests in space give results as good as those on the ground, these devices will prove to be practical. They might well become very attractive for propelling space probes (Figure 4.11) and, perhaps, even manned spacecraft.

An even more unconventional way to launch an interplanetary spaceship is to have, above the Earth's atmosphere, a rotating tether, a very long cable with two large masses at its ends, which rotates in a vertical plane while in orbit around the Earth. The direction of rotation is such that, when the tether is vertical, the peripheral velocity

Figure 4.11. A space probe sailing close to Jupiter propelled by the solar wind using a magnetic bubble (NASA image).

adds to the orbital velocity around the Earth at the higher end and is subtracted from it at the lower end, i.e. the higher end moves faster than the lower end. A payload is launched on a suborbital trajectory in such a way that it reaches the lower end of the tether at the instant it is vertical. If the velocities of the payload and of this end of the tether match exactly, the former can dock with the latter and can then be carried around in its motion. While going around attached to the tether, the payload gains speed and when it reaches the highest position it is travelling well above orbital velocity. At the top, the payload is released and tossed out into space. The payload can be injected into a rapid transfer orbit towards the Moon, Mars or one of the other planets.

Obviously the energy given to the payload comes from the kinetic energy of the tether, which slows down and moves into a lower orbit. After throwing the payload out into space the energy of the tether must be restored, which can be done using electric energy generated by solar panels; a current circulating in the tether can react with the Earth's magnetic field, generating a force which increases the rotational velocity of the tether. Alternatively, energy generated by the solar panels can operate winches, which change the tether length, transferring energy to the rotating tether (so-called gravity gradient propulsion). The advantage is that the spacecraft does not need to carry its own propulsion system. The tether, on the contrary, can be massive (the heavier the better, as it can store more energy and will slow down less in the tossing out process). Its propulsion system can be a low power one, as it can restore the velocity of the system gradually.

The rotating tether could also slow down an incoming payload, in which case it receives its kinetic energy. It can then be considered as a sort of energy accumulator, taking energy from the inbound spacecraft, storing it and giving it back later to outbound vehicles. If in the future there is considerable interplanetary traffic, rotating tethers in orbit around the Earth, the Moon and other planets could be very useful in avoiding the expenditure of large amounts of energy launching a variety of spacecraft and also receiving resources from space. As already seen for the other unconventional systems, such as skyhooks or space fountains, described in Chapter 2, such technological infrastructures are costly, and justified only when there are large volumes of traffic.

The ESA programmes Horizon 2000 and Post Horizon 2000 include a mission to Mars[5] and also one to the inner solar system, to study Mercury. NASA will concentrate its planetary exploration efforts on Mars, with a number of robotic missions, to be launched during the optimum launch windows of 2002, 2004 and 2006. In Japan, ISAS has a well-developed programme of robotic exploration of the Moon and the planets, with probes *Lunar A* aimed at studying the lunar surface and *Planet B* for the study of the atmosphere of Mars. Also, NASDA has some lunar missions in its programmes. Little is definite about future Russian programmes, but it is likely that, after the failure of *Mars 96*, the Russians will propose other missions, possibly an international one, to Mars. Mars Together is a provisional name for these missions.

In spite of the fact that the Moon has been visited by a number of astronauts, there is still much work to be done there using robotic probes. Before starting programmes aimed at the construction of a permanent outpost in the first decades of the twenty-first century, a number of automatic missions, including satellites around the Moon, robotic landing modules and devices to return lunar samples to the Earth, must be performed. As these activities are mainly in preparation for the construction of a lunar base, they will be dealt with in Chapter 5.

The next target for robotic exploration is Mars. The hope, bordering on certainty, of the existence of life (at least in the form of microorganisms) on the red planet was followed by disappointment as soon as the first probes reached Mars. The worst disillusion perhaps came in 1966; the images from the *Mariner 4* probe showed a desolate, lunar-like landscape. That was essentially a stroke of bad luck, since the small area depicted in those few images was exceptionally barren, even by Martian standards. But nobody knew that then. Even when

---

[5] *Mars Express*, to be launched in 2003, will carry the British *Beagle 2* lander.

the images of the *Mariner 9* arrived, hopes of finding life on Mars did not improve much.

Scientists preparing the *Viking* missions included on each four biological experiments, one of which was a gas chromatograph–mass spectrometer (GCMS) to examine soil specimens to search for organic molecules. There was also an experiment to identify the products of photosynthesis in the soil, a gas-exchange experiment and a radioactive isotope release experiment. Despite some initial results which seemed to be positive, both *Viking* landers showed that the soil does not contain any organic substance and there is no form of life on the surface of Mars nowadays. Ultraviolet and X-ray radiation from the Sun sterilises the surface. The reactions that at first suggested the release of gas from biological specimens were ascribed to reactions between inorganic matter and strongly oxidising substances present on the ground. Nevertheless, the absence of evidence for life on Mars is *not* equivalent to evidence for its absence.

The greatest mystery of whether life started on Mars more or less at the same time as it started on Earth is thus still open. Following the results of later probes which visited the red planet, we presume that the surface of Mars was, in the very distant past, rich in liquid water, so rich that clear erosion signatures in images of the Martian surface are still present. At that time, the Martian environment was probably not very different from the Earthly environment.

Since the experiments of *Viking* have recently been reinterpreted in a somewhat less negative light,[6] it may be that some primitive forms of life still exist on Mars. Moreover, life may well exist many metres below the surface, under special conditions, such as in a canyon like the Vallis Marineris, where the scarce atmospheric moisture could gather. Performing an automatic landing there is all but impossible. The most suitable zones for life on Mars can be reached only by vehicles, piloted or automatic, moving on the ground and able to manage the steep mountain slopes.

[6] Laurence Bergreen, *The Quest for Mars*, Harper Collins, London, 2000; M.Walter, *The Search for Life on Mars*, Allen & Unwin, St Leonards, New South Wales.

Figure 4.12. The ALH 84001 meteorite found in the Alan Hills region of the Antarctic in 1984 (NASA photo).

The discovery, made by a group of scientists led by David McKay in August 1996, that a meteorite which is supposed to have come from Mars contained something which was tentatively identified as fossil traces of lifeforms (Figures 4.12 and 4.13) caused much publicity and many arguments. The meteorite in question had been found in the Alan Hills region of the Antarctic in 1984 (as shown by its designation ALH 84001), and initially classified as a fragment of an asteroid, a diogenite. It was stored, with other meteorites, at the NASA Johnson Space Centre in Houston. It was only in 1993 that it was again studied and reclassified as an SNC meteorite. Since there are only 15 known meteorites of this type, ALH 84001 became a rare specimen, worthy of detailed study. When David McKay's team announced its results, some NASA spokesmen reported the news with supporting evidence, with the aim of bolstering funding requests for robotic missions to Mars and, ultimately, a manned mission.

There is little doubt that this meteorite actually came from Mars – the isotopic composition of the elements present in it is very

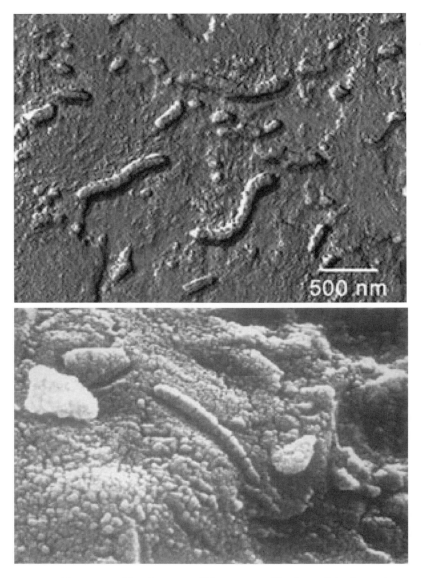

Figure 4.13. Two images at different magnifications of the formations which look like microfossils in an electron micrograph of the ALH 84001 meteorite; their length is less than a thousandth of a millimetre (NASA photos).

convincing proof of that. The meteorite could be of the SNC type, similar to the Shergotty meteorite found in India in 1865 (they are also called Shergottites) together with meteorites found in Nakhla (Egypt) and Chassigny (France). The designation SNC comes from the initials of these three localities.[7] All SNC meteorites are believed to have arrived on Earth from the red planet.

There is also little doubt that a fragment from the surface of a planet such as Mars can be ejected into space by the impact of a large meteoroid. It has even been possible to reconstruct in some detail the history of the ALH 84001 meteorite. It crystallised from molten rock about 4.5 billion years ago, and was then shocked by an impact some 0.5 billion years later. Sometime between 1.8 and 3.6 billion years ago, carbonate minerals were deposited in it by water flowing through the rock; at this time the formations which can be interpreted as the remains of living organisms were produced. Then 16 million years ago it was ejected into space by an asteroid impact, remaining there until it entered the Earth's atmosphere 13 thousand years ago to land in the Alan Hills region of Antarctica.

Contamination from terrestrial biological material seems to be ruled out, but the fact that the microscopic structures found inside the ALH 84001 meteorite are due to the activity of living beings is debatable. A biological origin is suggested by these structures looking like worms being similar to tiny fossils found on Earth; there are several clues, but no proof. Other researchers have found polycyclic aromatic hydrocarbons (PAH) in the same meteorite. These are often associated with life but are known also to be produced by non-biological reactions. Traces of iron sulphide, which often has a biological origin, were also found.

The microscopic structures are very similar to microfossils, but the evidence is not proof of a biological origin. Those who support this biological hypothesis assess that, even if no finding has a unique biological explanation, the coincidence of many different clues offers

---

[7] H.Y. McSween, Jr., *Meteorites and Their Parent Planets*, Second Edition, Cambridge University Press, Cambridge, 1999.

good evidence.[8] Much research work is still needed and it is likely that definitive answers can come only after experiments carried out directly on Mars.

Work on the Martian meteorite goes on; its scope has been widened to include other SNC meteorites, particularly the Nakhla and Shergotty meteorites, in which formations of the same type have been found. It is amazing that the ages of these two meteorites are so different. The first is about 1.3–0.7 billion years old, while the second is far more recent, just 165 million years old. If, therefore, these formations are really microfossils, Mars should have hosted living beings not only in the distant past, but also in relatively recent times, when dinosaurs were roaming the Earth.

Some scientists are convinced that, if life ever evolved on Mars, it could have survived to the present time in some particularly suitable places or in protected sites, such as in the permafrost layer which is thought to be present below the surface of most of the planet.

The possibility of finding traces of life on the red planet must not yet be ruled out altogether. It is a powerful reason for going on with the exploration of Mars. The *Sojourner* six-wheeled robotic vehicle transported by the *Pathfinder* probe which worked with much success did not seek to produce significant evidence on this issue. Future robotic missions, especially that to take samples of the surface and return them to the Earth for analysis (programmed for the launch window of 2007, or later), could yield extremely important results.

To explore the surface of Mars using mobile robots is not at all easy. First, communications from the controllers on Earth to the robot take about ten minutes or even more to arrive. Only when very slow operations are planned may movements of the vehicle be controlled from the Earth. When the robot encounters an obstacle, many minutes will pass before the controllers are aware of that and many more minutes before the correction commands arrive back at the

[8] A detailed discussion on this subject can be found in Bruce Jakosky, *The Search for Life on Other Planets*, Cambridge University Press, Cambridge, 1998, and Laurence Bergreen, *The Quest for Mars*, Harper Collins, London, 2000.

robot on Mars. The robot must therefore be able to work in quite an autonomous way, to decide the best path to follow and to modify that if necessary. Other difficulties are linked with the electrical power supply. Discounting nuclear generators and chemical fuels, which need an oxidiser to be carried from the Earth, since the Martian atmosphere does not contain oxygen in any useful quantity, the only possible choice is a system based on solar panels. At the distance from the Sun at which Mars is, their power output is considerably less than near the Earth and the robot can only come alive for a few hours each day. For these reasons the speed of the robot, its mass and its general mobility are rather small. The specifications (ROSA-M) for microrovers for Mars, issued by ESA, require a speed of just 5 metres per hour, about one fifth of the *Sojourner*'s speed.

The missions under way at the time of writing are sending back new information. *Mars Global Surveyor* discovered that the ice cap at the North Pole of the planet contains a quantity of water ice far larger than expected, evaluated to be more than a million cubic kilometres, i.e. half the ice present in Greenland. Some recent photos taken by the Mars Orbiting Camera (MOC) on *Mars Global Surveyor* have shown slopes carved by water streams. These formations are found in many places, particularly at high latitudes. They look geologically recent, and indicate that liquid water existed on Mars more recently than generally thought.

Another interesting instrument aboard *Mars Global Surveyor* is the Mars Orbiting Laser Altimeter (MOLA). It has obtained maps of almost all the surface of Mars with unbelievable precision. The data are crucial for the planning of future robotic and manned expeditions.

Unfortunately, *Mars Polar Lander*, launched in January 1999, was lost as it approached the South Pole of Mars in December 1999. Among many experiments it carried were two penetrators designed to reach a depth of about two metres, to study the temperature and composition of the soil.

An important step for the exploration of Mars will be a satellite in orbit around the planet to relay communications between probes on the surface and the Earth. The other probe launched in January 1999, *Mars Climate Orbiter*, after completing its meteorological tasks, should have remained in Mars orbit and acted as a telecommunications satellite. But it, too, crashed onto the Martian surface in September 1999.

The *2001 Mars Odyssey* probe, named to honour the (science fiction) author Arthur C. Clarke, was launched in April 2001 to Mars. The spacecraft was slowed down by rockets and, by aerobraking in October 2001, put into an orbit around the planet. In February 2002 it started mapping the amount and distribution of certain chemical elements, in particular hydrogen to search for water, and minerals that make up the Martian surface. It carries three main instruments – THEMIS (Thermal Emission Imaging System), to determine the distribution of minerals, particularly those that can only form in the presence of water, GRS (Gamma Ray Spectrometer), to study 20 chemical elements on the surface of Mars, including hydrogen in the shallow subsurface (i.e. water in the permafrost layer) and MARIE (Mars Radiation Environment Experiment). The main goals of the mission are to look for water, considered a prerequisite for the development of life, and to prepare the way for manned exploration. The orbiter will also act as a telecommunications satellite for future US and international landers.

The possible discovery of life on Mars, even if only fossil life, could have two opposite effects on the exploration of the planet. It might increase the number of scientific missions but, on the other hand, it could lead to the transformation of the planet into a sort of scientific park, delaying its exploitation and colonisation, discussed in Chapter 6.

NEW ROBOTIC PLANETARY PROBES

Apart from Mars, the only bodies in the solar system where there are some chances of finding life are some of the satellites of the giant planets. The *Galileo* probe sent back pictures of Europa, one of the satellites of Jupiter, which is slightly smaller than the Moon, having a

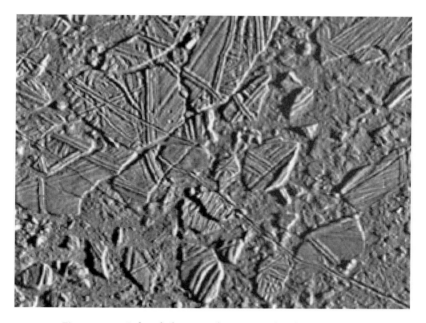

Figure 4.14. A detailed image of an area 34 km by 42 km on the surface of Europa taken by the *Galileo* orbiter (NASA photo).

radius of 1,565 km. These images could indicate that under its frozen surface there is an ocean of liquid water (Figure 4.14). The crustal plates are a few kilometres across; they have been broken apart, superficially resembling the disruption of pack-ice on polar seas during spring on Earth. The size and geometry of these features suggest motions of ice-crusted water or soft ice close to the surface at the time of their disruption. Measurements of the magnetic field have been interpreted to suggest that under a layer of ice there is an ocean of water. Neither the thickness of the ice nor the depth of the ocean are known reliably, however.

A possible schematic view of the interior of Europa is shown in Figure 4.15, derived from measurements of the gravitational and magnetic fields performed by the *Galileo* probe. The core (not sectioned in the figure) is surrounded by a shell of rock (sectioned), which is surrounded by a layer (darker in the figure) of water in liquid

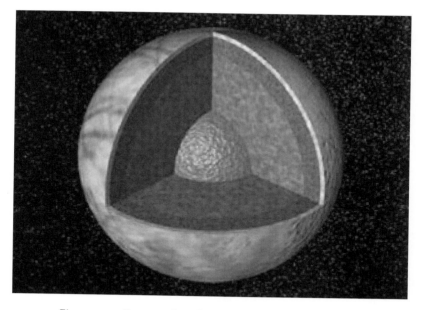

Figure 4.15. Cross section of Europa, with a sketch of its possible internal structure. For the surface, the images taken in 1979 by the *Voyager* probe have been used (NASA image).

form or ice. The external layer (in white) is surely ice, perhaps 10 km thick.

The possibility of the existence of liquid water at such large distances from the Sun, which until a few years ago were thought to be completely beyond the zone in which life could develop, should not really astonish us. A planetary satellite can be heated in many ways, such as by radioactive elements in its core, or by huge tidal forces due to the planet around which it orbits. The second reason is believed to be the reason why some satellites of Jupiter, for example Io, the most volcanic object in the solar system, are so very hot inside.

Europa is not the only satellite of the giant planets to be very rich in water. Ganymede, another of the Galilean satellites of Jupiter, may have about 15% of its mass as water ice, or as underground lakes. Life might exist, or even thrive, there. Regarding Io, protection from the strong radiation associated with Jupiter's charged particles is needed as Io's orbit lies well within the Jovian magnetosphere.

Titan, a satellite of Saturn, is more interesting from the point of view of searching for life elsewhere in the Universe. Titan is a large satellite, with a diameter of 5,150 km. It is one of the few satellites to have a rather dense atmosphere, twice as dense as that of the Earth. The atmosphere of Titan, studied in 1982 by the *Voyager* probe, hides the surface from direct observation. The low density of Titan suggests that large quantities of water are present, in the form of ice due to the low temperature. At the distance of Saturn from the Sun, the temperature of Titan's surface should be less than 100 K (−173 °C). The strong greenhouse effect due to its atmosphere raises the temperature somewhat, but according to the *Voyager* measurements it does not exceed −100 °C, even at high altitude. Titan will be studied in detail by the lander *Huygens* of the *Cassini* probe.

Another very interesting mission is *Pluto Express*, a robotic probe to be launched towards Pluto, currently planned by NASA. At present, the mission is in doubt, since the budget for Pluto exploration had been cancelled, but then reinstated, by the American Congress, and the Planetary Society and other organisations are trying to rescue it. Pluto and its very large satellite Charon (Figure 4.16), so large compared with the size of the planet and so close to it that many astronomers now define the Pluto–Charon system as a double planet,[9] are among the least-known bodies of the solar system.

Owing to the enormous distance of Pluto from the Sun, a large initial velocity will be used and, as a consequence, the trajectory of the spacecraft will be almost straight. This will be possible owing to the small mass of the spacecraft, thanks to the use of advanced technologies of the type incorporated in the *Clementine* probe. It has, however, the disadvantage that the probe will remain near the planet and its satellite Charon for only a very short time. It will be a 'fast flyby' mission, after which the probe will travel on towards interstellar space.

---

[9] The diameter of Pluto is 2,320 km, that of Charon 1,270 km, and the radius of the orbit of Charon around Pluto is 19,640 km.

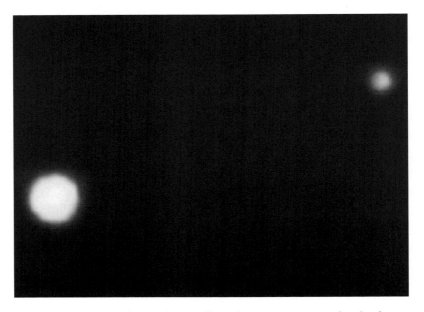

Figure 4.16. Pluto and its satellite Charon in an image taken by the ESA/ESOC Faint Object Camera mounted on the *Hubble Space Telescope* on February 21, 1994 (NASA photo).

The robotic planetary missions at present under study will use chemical rockets. The only probes with electric propulsion (solar electric) are those included in the New Millennium programme of the NASA Jet Propulsion Laboratory. That programme includes a number of deep-space missions, starting with the *Deep Space 1* probe, launched later than planned, in October 1998 (Figure 4.17). This mission is important not only for its scientific goals, but also because it is the first space test of an electric thruster for the main propulsion unit[10] of a spacecraft. The trajectory initially included the flyby of MacAuliffe asteroid and West–Kohoutek–Ikemura comet (Figure 4.18), but launch delays forced the objectives of the mission to be modified. At the time of writing the probe is on its planned course and the ion thrusters have worked satisfactorily.

[10] E.K. Kasani, J.F. Stocky, M.D. Rayman, *Solar Electric Propulsion*, First IAA Symposium on Realistic Near-Term Advanced Space Missions, Torino, June 1996.

Figure 4.17. The encounter of the *Deep Space 1* probe with the
West–Kohoutek–Ikemura comet (drawing from NASA Jet Propulsion
Laboratory).

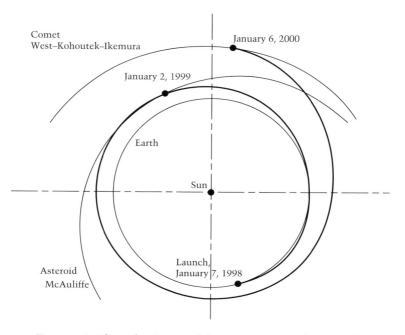

Figure 4.18. Planned trajectory of the Deep Space 1 probe toward
McAuliffe asteroid and West-Kohoutek-Ikemura comet.

EXPLORATION OF COMETS AND ASTEROIDS

Comets are another important target for robotic space missions, because they are fragments of the material which constituted the solar system at the time of its formation, before those processes which led to the formation of the planets started. The fact that on comets many amino acids, which constitute the basic building blocks on which life is based, have been detected makes their study even more important.

Comets are, unfortunately, difficult targets for space probes to reach. They cross the inner part of the solar system very quickly and, when they are close to the Sun, they move at very high speeds. When, at the beginning of the 1970s, planning for a mission to Halley's comet started, NASA thought it best to use an electric propulsion probe and then one propelled by a solar sail. The mission was then cancelled and the studies stopped. When Halley's comet entered the inner solar system in 1986 it was met by four probes, *Giotto* (European), *Vega* (Russian), *Sagigake* and *Susei* (Japanese). The *Giotto* mission, in particular, yielded a great deal of very interesting information, even though the speed at which the flyby occurred was almost 70 km/s (Figure 4.19). The time that the probe spent close to its target was thus quite short. The nucleus of Halley's comet is an extremely dark object; it reflects only 4% of the light incident upon it. By comparison, coal is less black; it reflects about 7% of the incident light. The pictures taken by the probe showed many details of its evaporating surface and allowed us to understand some of the basic mechanisms which control the life of comets.

Other cometary missions have been performed in recent years, such as the *International Comet Explorer* to Giacobini–Zinner and the second *Giotto* mission. For this the probe was awoken after a long hibernation and was directed, using the propellant saved from the previous encounter, towards the comet Grigg–Skjellerup. This time the flyby velocity was only 14 km/s.

A cometary mission called *Stardust* is at present under way. The spacecraft will perform a flyby of comet Wild 2 in January 2004. When at a short distance from its target, *Stardust* will capture particles

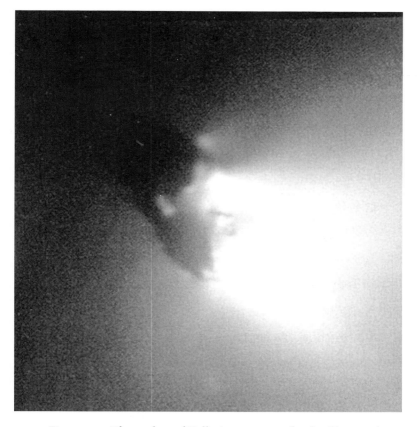

Figure 4.19. The nucleus of Halley's comet, seen by the *Giotto* probe from a distance of 4,000 km (ESA image).

detaching themselves from the nucleus of the comet using a silicon-based substance (aerogel). The samples collected from near the comet will be delivered to Earth in a capsule which will be jettisoned as *Stardust* passes near the Earth in January 2006. Also, samples of interstellar dust will be collected during the flight. These are interesting as they are expected to be ancient, pre-solar interstellar grains and other remnants from the formation of the solar system.

ESA is planning a new cometary mission, *Rosetta*. It is scheduled for launch in 2004 to reach a comet, perhaps Schwassmann–Wachmann 3 or Wirtanen, with a flyby of the Brita asteroid or to

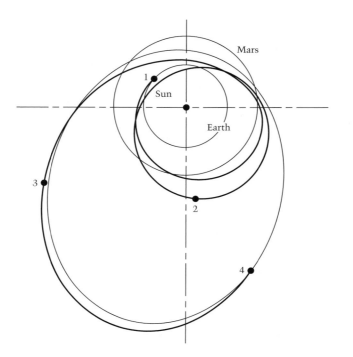

Figure 4.20. Trajectory of the *Rosetta* probe towards the comet
Wirtanen. 1: launch, 2: flyby of Mimistrobell asteroid, 3: flyby of Shipka
asteroid, 4: arrival.

Wirtanen, with a flyby of the asteroids Mimistrobell and Shipka
(Figure 4.20). The trajectory is quite complex with one Mars and two
Earth flybys to increase the probe's speed. The probe will have modules
which will land on the surface of the comet, and analyse samples of
it. The original plan was to bring back to Earth some samples of the
solid material of the comet, but that part of the mission was cancelled
owing to its difficulty and cost. The mission will thus be limited to
analysis on the spot of some specimens and to broadcasting the results
back to the Earth.

Asteroids are another class of interesting targets for robotic
missions. Some of them are relatively easy targets, as their orbits
cross the Earth's orbit or, at least, come quite close to it. Some of
these asteroids could be a potential source of danger for our planet

and should be the object of detailed study in the near future. While the probability of an impact of a small asteroid or cometary nucleus with the Earth is small, its consequences would be so dramatic that monitoring is required. The impact of the comet Shoemaker–Levy 9 on Jupiter[11] has attracted public concern to such events which, while being very rare on a historical timescale, have nevertheless occurred several times in the past. The theory that mass extinctions,[12] which several times in the geological history of our planet wiped out a large number of species (up to 90% in some cases), are due to large meteorites crashing to the Earth's surface and putting huge quantities of dust into the atmosphere is gaining popularity.

The falling of a smaller object, a comet nucleus rather than a meteorite like that causing the Tunguska event in Siberia in 1908, is more likely. It would not produce effects on a global scale, but would cause a large loss of life and great damage if it fell on a populated area.

The seriousness of the impacts of meteoroids, asteroids and comets has been codified in a scale, called the Torino scale. Conceived in 1995 and officially accepted at a conference held in Turin in 1999, it is similar to the Richter scale for earthquakes. The danger is represented by 11 points (from 0, no risk, to 10, certainty of general destruction) and five colours (from white, certainty of no impact, to red, certainty of impact).

When an asteroid is discovered, its orbit can be calculated only in an approximate way. Then the potential danger which it represents may be assessed only in statistical terms. Even when the orbit becomes better known, it can be changed by gravitational perturbations due to the planets or even the Earth in such a way that the danger changes. These perturbations cannot be computed with the required precision – a variation of a few thousand kilometres, a trifle on an

[11] J. Crovisier and T. Encrenaz, *Comet Science: The Study of Remnants from the Birth of the Solar System*, Cambridge University Press, Cambridge, 2000.
[12] V. Courtillot, *Evolutionary Catastrophes: The Science of Mass Extinction*, Cambridge University Press, Cambridge, 1999.

astronomical scale, may transform a harmless asteroid into a serious danger for human beings on planet Earth.

Up until July 1999, 803 asteroids whose orbits approach that of the Earth (NEA, Near Earth Asteroids) were known, 304 of which have a diameter equal to, or larger than, one kilometre. These could cause a global catastrophe. One hundred and eighty three asteroids have been classified as potentially dangerous (PHA, Potentially Hazardous Asteroids), as they could come dangerously close to the Earth. None of the known objects has a degree higher than 1 on the Torino scale.

The asteroids which come close to the Earth are interesting as potential sources of raw materials when the time to exploit space resources arrives. Mining dangerous asteroids to obtain their raw materials is the safest way of avoiding the danger of their collision with us.

The recently founded Space Guard Society plans to monitor space using telescopes to identify these small, yet potentially dangerous, objects. In the future a network of satellites could perform this task. What should we do when an object is found which could cause significant damage if it were to strike the Earth? To blast the meteoroid into pieces without changing its trajectory would be harmful – it would trade a large object for a swarm of small fragments, making things worse. The incoming asteroid or meteoroid must be deflected slightly when the object is still far from the Earth, but there is no agreement on the suitability of using nuclear missiles for that purpose.

The *NEAR* (Near-Earth Asteroid Rendezvous)[13] *Shoemaker* (after the name of the famous comet discoverer) probe made a close flyby of the asteroid Mathilde (Figure 4.21), a C-type asteroid with a diameter of only 33 km, passing it at a distance of only 1,200 km in June 1997. The study of the 534 images taken and the readings of the instruments yielded much information on such asteroids, which are

[13] Factual information on the *NEAR* mission, and indeed on every space mission launched during the twentieth century, is available in R. Zimmermann, *The Chronological Encyclopaedia of Discoveries in Space*, Oryx Press, Phoenix, Arizona, 2000.

Figure 4.21. Pictures of three asteroids which have been visited by automatic space probes. From the left, Gaspra and Ida, silicon-rich asteroids visited by *Galileo* and, on the right, Mathilde, seen by *NEAR* (NASA photos).

thought to be samples of the 'original stuff' of the solar system when the planets were formed. They are called C-type asteroids because they are very rich in carbon. After that encounter *NEAR* was pointed at the main target of the mission, the asteroid Eros. This is a member of the class of bodies crossing the orbit of our planet.

NEAR reached Eros and began to orbit the asteroid in February 2000. For almost a year it took 150,000 photos, about ten times as many as originally planned, changing its orbit from time to time to study its target from different altitudes. The surface of the rocky asteroid has now been mapped in detail, and all important features have been given names. One such image is shown in Figure 4.22: the bright, dome-shaped features are natural, not structures built by aliens as some have suggested!

At the end of the spacecraft's useful life and with nothing to lose, the mission engineers decided to attempt to land the probe on Eros, in spite of the fact that it was not designed to perform a landing. Descent to the surface took place in February 2001; 69 images were taken before touchdown, the last of which is shown in Figure 4.23. The spacecraft survived the landing and some instruments continued

Figure 4.22. A close up picture of the Eros asteroid taken by the *NEAR* probe (NASA photo).

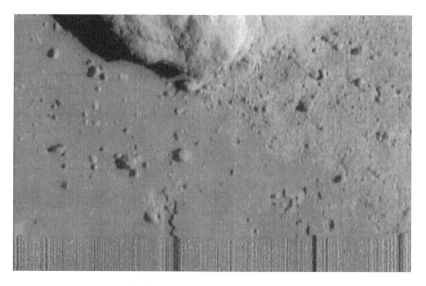

Figure 4.23. The last image of Eros taken by *NEAR* before touchdown, from a distance of 120 m; the width of the zone imaged is 6 m. The boundary between rough and smooth terrain running diagonally across the picture is unexplained. In the lower part the comb-like appearance is due to loss of the radio signal at touchdown (NASA photo).

working for a further 14 days. *NEAR* is the first object built by humankind to land on an asteroid.

Missions to the asteroid belt, located between the orbits of Mars and Jupiter around the Sun, will allow the study of several different types of asteroids. A target of particular importance is Vesta, a large asteroid which in the past was hit by a smaller one. That caused a crater deep enough to expose the material which constitutes its core – the core of an asteroid can thus be studied without having to dig deep into it.

THE KUIPER BELT AND THE HELIOPAUSE

The Kuiper belt is a zone beyond the orbit of Neptune in which many solid bodies similar to large asteroids have been identified. Its distance is so great that only its larger bodies can be identified at all. As our observational techniques advance, it is becoming clear that it is a true second asteroid belt, far larger than that located between Mars and Jupiter. Pluto and its satellite Charon are now thought to be representative of the larger bodies in the Kuiper belt, and this can explain the anomalies of Pluto's orbit.

The thickness of the Kuiper belt is unknown. Some astronomers think that it may extend out to the Oort cloud, the spherical region around the Sun whose radius is about 50,000 times the Sun–Earth distance, consisting of blocks of ice of varying shapes and sizes. Some of these move towards the inner solar system, because of perturbations to their orbital motions due to other stars, with very elongated orbits. They become comets. Thus, the Oort cloud is the place from which comets come. They are made of material which has remained unchanged since the formation of the solar system.

Some bodies which have a coating of ice like comets, but with an inner core made of rocks like an asteroid, have been identified in the Kuiper belt. They have been called 'cometoids', and their presence provides evidence that the outer portion of the Kuiper belt reaches out to the Oort cloud. It has been suggested[14] that some of the bodies in the

[14] R.L. Forward, Alien life between here and the stars, in *Islands in the Sky*, Wiley, New York, 1996.

Kuiper belt may host lifeforms, obviously quite different from those we know. Some scenarios regarding their development have been proposed. A probe like *Pluto Express* could be directed, after its rapid flyby of Pluto, towards an asteroid or cometoid in the Kuiper belt and make important observations there.

The Sun moves, with the planets and all the bodies of the solar system, through that very rarefied gas which constitutes the interstellar medium. In its motion through the interstellar gas, the interplanetary medium creates a shock wave ahead of it, similar to what happens when a supersonic aircraft flies through the air. The distance of this shock wave from the Sun is not well known, but it is clear that it has to be much closer to it in the direction of the motion of the Sun. The shock wave is well beyond the orbit of Neptune and well within the Oort cloud, which leaves a span of thousands of billions of kilometres. It is the true frontier of the solar system, marking the separation between the interstellar medium and the medium which fills our solar system; it is referred to as the heliopause. A picture of the heliosphere is shown in Figure 4.24.

The two *Voyager* probes, which are now well beyond the orbit of Neptune, have not yet reached the heliopause and it is possible that they will not reach it with their instrumentation and radio transmitters still operating. Missions to investigate the heliopause can be performed using chemical rockets, but to avoid journeys of several decades it would be advisable to use electric propulsion or solar sails.

At the beginning of the 1980s NASA's Jet Propulsion Laboratory suggested the *Tau* (Thousand Astronomical Units) mission, to reach a distance of 1,000 Astronomical Units from the Earth, i.e. a distance equal to a thousand times the Sun–Earth distance, or about 150 billion km. The mission had to be performed using ion thrusters, powered by a nuclear reactor. The project was then abandoned, even though a mission to reach the heliopause has scientific appeal, and the use of electric thrusters for such a long time and the transmission of data from such a huge distance are in themselves important technological challenges.

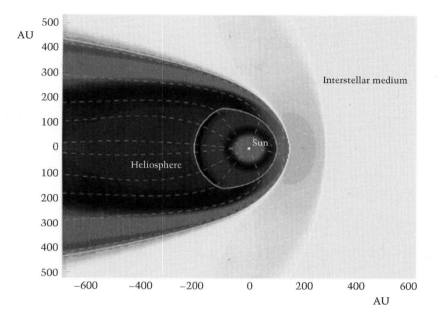

Figure 4.24. Diagram of the heliosphere (the dark region), with distances shown in Astronomical Units.

Recently, an international group has proposed a solar sail probe, named *Aurora*,[15] to exit the solar system at high speed. After leaving the Earth's sphere of influence, the sail would first be oriented to slow the probe down, so that it starts falling towards the Sun. When the perihelion (the point of minimum distance from the Sun) is reached, at about 0.2 Astronomical Units, the sail would be angled so as to exploit the 25 times greater (than at the Earth's distance from the Sun) radiation pressure to accelerate the probe towards the outer solar system. *Aurora* should leave the solar system at a speed (relative to the Sun) of about 11 Astronomical Units per year, or about 52 km/s. For comparison, the speed which the *Voyager* probes had when crossing the orbit of Neptune was about 16 km/s.

[15] Not to be confused with ESA's long-term strategy for planetary exploration, also termed Aurora.

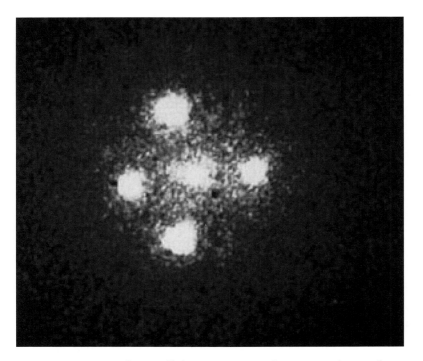

Figure 4.25. The so-called 'Einstein Cross', four images of a very distant quasar multiply-imaged by a relatively nearby galaxy acting as a gravitational lens, taken by the ESA Faint Object Camera aboard the *Hubble Space Telescope* (NASA photo).

THE FOCAL LINE OF THE SUN'S GRAVITATIONAL LENS

Among the results of the theory of relativity there is the prediction that an extremely massive body bends beams of light, owing to its gravitational attraction, through a small angle. The Sun, or any other star, acts on beams of light as a lens does in classical optics. This is usually referred to as a 'gravitational lens'. But there is a difference. While a perfect lens makes the light beams converge to a single point, the focal point, a gravitational lens makes them converge to a number of points, located on a straight line, the focal line. This effect has been observed in images of very distant astronomical objects such as quasars, the gravitational lens effect being produced by a comparatively nearby galaxy (Figure 4.25).

If a space probe were to be located in the focal line of the gravitational lens of the Sun, its instruments could make otherwise impossible observations. For example, the galactic centre, where a black hole might be located, could be studied in detail. The gravitational lens of the Sun focuses not only visible light onto its focal line, but all kinds of electromagnetic waves, including radio waves. People seeking the existence of intelligent extraterrestrial beings by searching for the radio waves which they might broadcast are very interested in a space mission carrying a receiving antenna to the focal line.

This line starts far outside the solar system, at 550 Astronomical Units (about 82 billion km) from the Sun, and continues outwards for billions of kilometres. The probe need not be stopped when it reaches this distance, as it will remain on the focal line for tens of years in its voyage through interstellar space. Recently, doubts have been expressed on using the gravitational lens effect for astronomical work and, particularly, for the Search for ExtraTerrestrial Intelligence (SETI). Turbulence and geometrical asymmetries of the solar corona could cause low quality 'images', i.e. the Sun is a poor lens. Further, it is almost impossible to aim the 'telescope' made by the Sun and the spacecraft in the required direction.[16]

Such a difficult mission has been proposed to combine astronomical studies and the search for extraterrestrial intelligence with studies of the heliopause, the interstellar medium and the Kuiper belt, but the first two goals are highly controversial. It is impossible to use a telescope with such a high magnification to search for something whose position is not precisely known, or whose existence is still unknown. The use of electric propulsion or of a solar sail seems to be a must, the time to reach the beginning of the focal line with present technologies being about 30 years.

[16] G. Genta and G. Vulpetti, Some considerations on Sun gravitational lens missions, *Journal of the British Interplanetary Society*, Vol. 55(3/4), pages 131–136, 2002.

5   Back to the Moon

Should the planning and construction of a permanently occupied base on the Moon be the next major objective for the human species in space? Would it be used best for scientific studies, e.g., of lunar geology, or of optical or radio astronomy? Or for demonstrating new technologies such as the extraction of mineral resources? Or as the site for a power station? Whichever is chosen, it would clearly show the potential for our human species beyond the Earth. What difficulties and hazards would be experienced? What would be the best base layout, and how long would it be before it could be finished? And should a human colony on the Moon be the next step along our path to the final frontier?

SHOULD WE RETURN TO THE MOON
OR GO STRAIGHT TO MARS?

Back to the Moon is a goal and also a popular slogan in the aerospace community. At meetings of the International Academy of Astronautics it can be seen written on the pins proudly worn by some members. After the success of the *Apollo* missions (Figure 5.1), many thought that the colonisation of the Moon was at hand. The reasons why that did not happen have already been discussed.

Why is the movement for our returning to the Moon now regaining momentum, and is it worthwhile to work in this direction? In times of limited budgets, is it not better to concentrate our efforts and to aim directly towards Mars? The proposal for a permanent base on the Moon will, without doubt, cause arguments similar to those seen about the Space Shuttle and the *International*

Figure 5.1. The Lunar Excursion Module takes off, an image of humankind abandoning the Moon – the drawing refers to the *Apollo 11* mission (NASA image).

Space Station, and those who fear that such a project will take resources away from scientific missions and robotic probes will be against it. The only answer to this opposition is the positive statement that, if humankind is to expand in space, the presence of more than a few astronauts and cosmonauts in space at any one time is essential.

But among those who agree with this last statement there is no consensus about the construction of a base on the Moon. To simplify the situation, it is possible to identify three 'parties' among those who advocate a permanent human presence in space: the Moon party, the Mars party and the 'in space' party. Members of the first party claim that, since the Moon is the celestial body nearest to us, it is the natural candidate for our first experience of colonisation. They insist that the exploitation of lunar resources is a good reason in itself. The Moon is, in their opinion, the first place beyond the Earth where

Figure 5.2. A lunar city built in some small craters; larger flexible structures attached to the rim of the craters maintain a breathable atmosphere around the buildings (*Tent City on the Moon*, painting by Kazuaki Iwasaki).

humankind can live, work, create cities and prosper both culturally and economically (Figure 5.2).

Those who support the idea that humans should concentrate directly on the exploration of Mars claim that the Moon, with its lack of atmosphere, low gravity and a day that is almost one month long, is too unfriendly a world for human colonisation. Mars, on the other hand, is a true planet, which could be made suitable for human life. Those who support the colonisation of space itself consider that the future of humankind is not on the planets but directly in space, namely in large space habitats.

Actually these three ways of seeing the future of humankind are not so different from each other as they may seem at first glance, at least as far as their long-term goals are concerned; all three contain very reasonable aspects. The difference is more in the early priorities than in the ultimate goals, as it is likely that humankind will eventually settle on both the Moon and Mars,[1] as well as on many other celestial bodies, and that many human beings will live permanently

[1] See also P. Bond, *Zero g: Life and Survival in Space*, Chapter 11, Cassel, London, 1999.

in space habitats. The difficulty is in making the choice of a few pro-grammes on which to concentrate our scarce resources, in order to avoid wasting money and cancelling projects already started. Whilst the proposal for a permanent base on the Moon can be considered as a distraction by those who advocate settlements on Mars in the short term, the latter may be considered too ambitious. If it failed, that would set back all space programmes.

One objection to settling the Moon is that the speed of a rocket needed to reach the Moon is only marginally smaller than that needed to reach Mars, and that the difficulties of reaching both celestial bodies are not dissimilar from the viewpoint of the propulsion sys-tem. On reaching Mars it is possible to aerobrake using the planetary atmosphere while near the Moon all manoeuvres must be performed using rocket engines; thus, some argue that it is simpler to travel to Mars than to the Moon.

However, there are many factors other than propulsion which must be considered. Distance is an issue, the Moon being roughly a thousand times closer to the Earth than Mars. Radio communications with Mars are therefore much more difficult than with the Moon. The simple exchange of messages with people on Mars would take many minutes. Any emergency could become critical in that time – telemedicine is very difficult and severe psychological problems might result. The travel time to Mars is about one hundred times longer than to the Moon, which makes it necessary to take more care about protecting the crew against cosmic rays and charged-particle radiation from the Sun. A spacecraft designed for the Mars route must have a shelter in which the crew can be protected from solar flares. That is not essential to reach the Moon, as was demonstrated by the *Apollo* missions. Observation from spacecraft like *Soho*, which can give early warnings about the Sun's activity, can make our way to the Moon even safer. Also, the long period of weightlessness could be detrimental to the crew, and many projects for Mars missions include severe exercise regimes or artificial gravity produced by rotation of the spacecraft. Further, the need to stick strictly to narrow launch windows – which occur once every two years or so – for energy efficiency considerations, makes it far more difficult to organise a rescue expedition to Mars than to the Moon.

The exploration of Mars and the construction of outposts on that planet involve more complex expeditions, with longer durations and larger crews who would have to operate under far more difficult conditions. The costs and the risks of failures will increase, and a disappointing outcome or, worse still, the loss of human life, would probably set back exploration missions for decades.

The role of the Moon in the space option is thus very important, as it must eventually supply the raw materials for the construction of large orbiting structures such as space colonies and solar power stations. Mars could never play that role, both because of the greater distance from the Earth and for its stronger gravity; the resources on Mars will be useful for the colonisation of the planet itself, but it is unlikely that there will be anything on Mars worth importing to the Earth.

Going back to the Moon is, then, not just an option, but a compulsory step if humankind wishes to expand into space. And, even though the number of people who would settle on the Moon will be small, it is likely that the colonisation of space and of Mars will need logistic support from lunar outposts.

Ben Finney[2] noted that the sea-faring peoples of Polynesia acquired their navigation skills because they came from a land which had many islands not too far from its shores. To have a large celestial body, the Moon, so close to its home planet could have the same importance for humankind.

THE RATIONALE FOR SCIENTIFIC MISSIONS
ON THE MOON

The six *Apollo* missions, which landed a total of 12 men on the surface of the Moon, yielded first-class scientific results. One of the problems which seems to have been solved is that of the origin of the Moon; it is now considered to be a part of our planet detached from it by a collision

[2] B.R. Finney and E.M. Jones (editors) *Interstellar Migration and the Human Experience*, University of California Press, Berkeley, California, 1985.

Figure 5.3. A lunar rock sample (measuring 18 × 20 × 25 centimetres and weighing 6.4 kg) carried back by the *Apollo 17* astronauts from the Taurus–Littrow landing site (NASA photo).

with a very large asteroid. The isotopic composition of the specimens brought back by the astronauts seems to prove that beyond reasonable doubt (Figure 5.3). But, as already stated when dealing with the *Apollo* programme, those six missions could answer only in a limited way other scientific questions regarding the Moon. And it could not be otherwise; it is scarcely possible to know all about a celestial body after a few stays of a few hours each.

The lack of a lunar atmosphere has its advantage. The surface has remained practically unchanged, except for the impacts of meteorites of all sizes, and samples can be collected which have been there since the Moon was formed. Fragments of meteorites can be found which did not undergo the fierce heating due to atmospheric entry, as occurs in the Earth's case. The so-called seas of the Moon (maria) are large areas covered by molten rocks coming from the inner regions of our satellite which came to the surface as a consequence of the impact of some very large meteorites. The study of these volcanic rocks is very interesting.[3]

One of the most important issues regarding the Moon is the presence – or absence – of water, in the form of ice, there. Even if recently the *Clementine* and *Lunar Prospector* probes seem to have found hints of its presence under the surface in the polar regions, the question of how much water is present remains open. There might be enough water to fill a lake of 10 square kilometres to a depth of 10 meters. However, the water ice might be contaminated with frozen carbon dioxide and other solidified gases.

Since it is far easier to obtain oxygen from water than from the lunar rocks, which nonetheless contain large quantities of oxygen, the presence of water would make it attractive to establish human settlements on the Moon. Oxygen is important not only for breathing in and directly supporting human life, but also as an oxidiser for rocket engines. Hydrogen, which can be used as fuel, can be readily obtained from water by electrolysis. Lunar dust and rocks are rich in silicon and oxygen, with a very small but non-negligible quantity of hydrogen (about 50 parts per million, ppm) and helium (10 ppm), implanted by the solar wind. A certain amount of carbon should also be available, from chondritic meteorites.

Regolith, the layer of lunar dust which covers practically everything on the Moon's surface, proved to be much thinner and therefore much less dangerous than was feared by some before the *Apollo* missions. The lunar surface can support the weight of people and their equipment, and astronauts can lope around with much agility. In the *Lunar Rover* (Moon Buggy), an electric car with four wheels and a weight of 36 kg on the Moon, the *Apollo 17* astronauts travelled to a point 37 km from the landing spot, reaching a top speed of 20 km/hr (Figure 5.4).

The study of lunar geology and the assessment of lunar resources will be amongst the most important activities to be performed on the

---

[3] G.H. Heiken, D.T. Varinar and B.N. French (editors), *Lunar Sourcebook: A User's Guide to the Moon*, Cambridge University Press, Cambridge, 1991.

Figure 5.4. *Apollo 16* astronauts explore the Moon's surface with the *Lunar Rover* in April 1972 (NASA photo).

Moon. ESA issued a proposal for a lunar exploration programme with the eventual aim of constructing a manned outpost, with three preliminary phases – the launch of automatic probes, a permanent robotic presence and a start to exploiting lunar resources. Also, NASDA plans a number of missions to observe the lunar surface, first from space and then using automatic rovers, and finally to take and return samples to the Earth. After 2010, construction of the necessary infrastructures would start and a manned scientific outpost could be completed by 2030.

One proposal for a lunar mission would put an automatic probe of about 1,000 kg at the Peak of Eternal Light, the highest point of the rim of the crater in which the South Pole of the Moon is located. The peak and part of the rim of the crater are continuously illuminated by the Sun, making energy production via solar panels possible without interruption, while inside the crater, always in the shade, some volatile substances (among which should be water) could be exploited.

The idea of a competition and sponsors was floated for this mission. Will the future see spacecraft and robots covered by advertising trademarks like today's Formula 1 racing cars, and astronauts with spacesuits in which the logo of the space agency is lost among other logos and commercial icons? The Russians started this trend by selling broadcasting spots from the *Mir* station. This can only be a welcome innovation if it generates new funds and hastens the time when human beings will live on the Moon.

It is probable that the first commercial exploitation of the Moon will be linked with the entertainment industry, and the first lunar resource to be exploited will be its beautiful landscape. The idea is simple and does not require very large investments – a medium-sized rover with TV cameras is guided from the Earth. A simplified and smaller version of the *Lunar Rover* used by the *Apollo* astronauts or the Russian *Marsokhod* (Figure 5.5) would be more than sufficient.

Figure 5.5. A Russian rover like the *Marsokhod* shown on the lunar surface (NASA digital artwork by Boris Rabin).

Images taken by the vehicle and received by television sets could perfect the illusion of being on a vehicle travelling on the Moon; the virtual tourists could even drive the vehicle.

Such commercial undertakings could have a strong impact on lunar exploration programmes. The Moon would feature within the range of interests of a growing number of people, and the costs of lunar missions would come down as small-scale mass-production techniques were introduced.

From the viewpoint of scientific research, the Moon has a very important feature. Its far side, which never faces the Earth, is the only place in the solar system which is completely immune from disturbances due to human activity. This is important for radioastronomy or for optical astronomy, since on the Earth it is increasingly difficult to find sites suitable for large telescopes owing to the diffuse artificial light, especially street lighting. For optical, ultraviolet and near-infrared astronomy, large orbiting telescopes near the Earth, such as the *Hubble Space Telescope*, offer advantages over Earth-based telescopes. First and foremost, they are above the Earth's absorbing and twinkling atmosphere. The microgravity environment in Earth orbit allows the use of very large mirrors without any deformation due to their weight and vibrations due to seismic activity. A telescope, particularly a very powerful one, is so sensitive to the motion of its support as to become completely useless if operating in the vicinity of human activities which produce the slightest vibration of the ground.

On the other hand, a telescope on the surface of the Moon, located at a reasonable distance from a permanent base, could be operated directly from there. The experience gathered running the *Hubble Space Telescope* has proved that maintenance and instrument changes are required from time to time. Although possible, such operations are complex and costly. If such an astronomical observatory on the Moon were to be built using materials extracted directly on the spot, it would not require the expensive launch of large quantities of material from Earth.

The Moon is not seismically active, an important feature for the construction of telescopes and large interferometers – arrays of small telescopes whose light is combined in such a way that collectively they work as a single telescope whose size is as large as the array. The relative positions of the individual instruments must be retained with a precision of the order of the wavelength of light. Such a feat is very difficult, not only in the case of telescopes 'free flying' in space, but also on the surface of a geologically active planet like the Earth. A large Moon-based interferometer could study extrasolar planetary systems some tens of light years from us in comparable detail to our knowledge of the solar system before the space age. Astronomical lunar bases can thus be a great asset, but only in the fairly distant future.

In the case of radioastronomy, on the other hand, operation in Earth orbit does not provide an environment free from human interference. Even if radioastronomical satellites, such as the Japanese *Haruka*, already in operation, or the Russian *Radioastron*, presently under study, can be very good scientific instruments, increasing space activities and dramatically increasing telephone and television traffic via telecommunications satellites will make things worse in the future.

If the problem of radio pollution (interference) is important for radioastronomy in general, it becomes a very bad nuisance for that branch of this science which deals with extremely weak signals, which can be completely lost in the background noise. Jean Heidmann,[4] for instance, believed that within a few years the search for extraterrestrial intelligence performed using radiotelescopes will become altogether impossible from the Earth or from near-Earth space, owing to many types of interference produced by humankind.

An astronomical and radioastronomical base on the far side of the Moon (Figure 5.6) could be an ideal solution, provided that in the meantime humans do not pollute this still-pristine environment with radio waves. Some radioastronomers proposed an agreement between the spacefaring nations to ban any form of radio transmission from

[4] J. Heidmann, *A New IAA Cosmic Study: Establishing a Radio Observatory on the Moon's Farside*, 50th International Astronautical Congress, Amsterdam, October 1999.

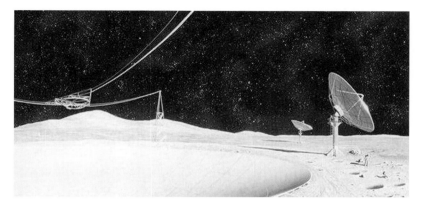

Figure 5.6. A radiotelescope of the same type as that at Arecibo, Puerto Rico, built in a lunar crater (NASA image).

artificial satellites of the Moon and lunar bases in a certain area. The most suitable place for an observatory seems to be the Saha crater, located slightly South of the Moon's equator at a longitude of about 102° E (Figure 5.7). It is a circular crater, with a diameter of about 100 km, surrounded by mountains 3000 m high, protecting the inner surface from short-wave radio transmissions from the surface of the Moon or from space vehicles not too high in the sky. It is never in sight of the Earth. But radio commands from the Earth could be sent to a station in Mare Smithii and a permanent connection established through a cable 350 km long or via three laser relay stations.[5] Data from the observatory would be relayed back along the same route.

Initially, it would be an automatic station. A single space vehicle, containing five landers, could be launched from the Earth. The largest lander would carry the radiotelescope to land in the Saha crater, and three small landers, each carrying a laser relay station, would land at the locations marked 1, 2 and 3 on the map (Figure 5.7). The fifth lander would be a radio station, to land in the Mare Smithii, to allow radio communications with the Earth. The radio astronomy station on the

[5] G. Genta, Twin rigid-frames hexapod rovers for the Saha radioastronomic missions, *Advances in Space Research*, Vol. 26, No. 2, pages 351–357, 2000.

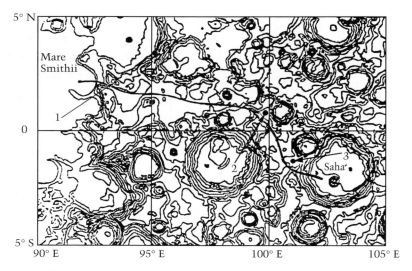

Figure 5.7. Map of the lunar territory including the Saha crater and the Eastern end of the Mare Smithii. The continuous line is the shortest 'road' connecting these two places; the points marked 1, 2 and 3 are the sites where three laser relay stations could be located.

far side of the Moon may subsequently be enlarged. When a permanent lunar base is built, it would be possible to transform what was initially a small automatic station into a fully fledged observatory. If not permanently manned, it would at least be visited from time to time by astronauts and scientists. They would perform all the necessary maintenance, and upgrade operations, e.g. by installing new equipment.

LUNAR OUTPOSTS

When manned missions to the Moon are resumed, much longer stays than those of the *Apollo* programme must be scheduled, to open the way to a permanent settlement on the Moon. Longer stays are made mandatory by the high travel costs and by the need to maximise the scientific return of each journey. It is impossible to have the crew living for more than a few days in the cramped space of the landing spacecraft, and so there must be some sort of pressurised shelter granting a minimum of comfort.

Figure 5.8. A module (similar to those built for the *International Space Station*) is shown being unloaded from a Moon lander onto a lunar transporter, before becoming part of a lunar base for humans, in this artist's impression (NASA image).

Initially, a lunar outpost could be made with habitation modules and arrays of solar panels very similar to those of the *International Space Station*, placed on the lunar surface and partially buried in it (Figure 5.8). Covering the modules with lunar dust provides thermal insulation and some protection against radiation. Technically, the construction of such a settlement is not difficult. The various modules can be launched into low Earth orbit in the same way as those which form the space station. The cost of building a few additional space station modules should not be too high, since the hardware launched into space represents only a small fraction of the overall cost of a space mission, with the design and development expenses representing a far larger part. To launch the modules into low Earth orbit would be the greatest expense.

The transfer of the modules from low Earth orbit to a low orbit around the Moon can be performed at a relatively low cost. To reach an orbit around the Moon from LEO is slightly easier than to reach a geostationary orbit.[6] The reusable Orbital Transfer Vehicle which was planned for the space station could lead to an *Apollo*-style manned expedition. Since during the transfer phase the lunar base modules would be unmanned, long transfer times (about 200 days) would be possible, using electric propulsion or chemical rockets, or even solar sails. The trajectories could exploit the 'fuzzy boundary'[7] between the gravitational fields of the Earth, the Sun and the Moon.

The descent to the Moon is the only part of the transfer which requires a strong thrust, in order to slow down the module's fall onto the lunar surface. A chemical rocket is needed and, as its size cannot be small, this phase cannot be performed at low cost. Once on the Moon, the modules must be connected to each other and then buried in the ground. Both operations should not be too difficult, particularly if a sandy location has been chosen, thanks to the reduced gravitation of the Moon. The simple shelter to house the astronauts during their stay will thus quickly become a true outpost, which could be useful not only for other scientific tasks, but also to experiment with the technologies needed to establish a permanent human presence on the Moon.

The volcanic nature of part of the Moon's surface suggests the possibility of building an even more economical and more comfortable shelter.[8] The lava flows which formed the 'maria' (the parts which, appearing darker to the observer, were first thought to be seas) are likely to have lava tubes, i.e. subsurface tunnels of different sizes (up to several metres diameter). If the same phenomenon occurred on the Moon as on the Earth, the astronauts might find large caves there, well protected from cosmic radiation and temperature changes, and ready to be transformed into comfortable dwellings. There are projects afoot to send robotic probes to find suitable tubes close enough to the

[6] R.C. Parkinson, *Citizens of the Sky*, 2100 Ltd., Stotfold, 1987, page 91.
[7] See Appendix B.
[8] Details are discussed in P. Eckart (editor), *The Lunar Base Handbook: An Introduction to Lunar Base Design, Development and Operations*, McGraw-Hill, New York, 1999.

surface. A vertical shaft can then be drilled, the inner surface made airtight and an airlock installed.

PERMANENT BASES

With time, the outpost could be enlarged by adding other modules such as laboratories, stores, workshops and living quarters for additional personnel. It could expand both on the surface and under it. Further living space could be obtained by using technologies more typical of civil engineering than aerospace engineering, using the range of materials found on the Moon.

Lunar rocks can make a very good concrete: samples have been made by grinding some small rocks carried back by the *Apollo* astronauts. The strength of the lunar concrete was found to be slightly higher than if normal construction materials on Earth had been used.

The primary objectives of a lunar base (following Koelle[9]) are to:

- Provide a science laboratory in the unique environment of the Moon.
- Improve our knowledge of the Moon and its resources.
- Develop marketable services and space products on the Moon.
- Establish the first extraterrestrial human settlement.
- Supply space-based energy to the Earth.
- Provide a focus for the development of space technology.
- Demonstrate the potential for growth beyond the Earth.
- Enhance the evolution of human culture in space.
- Provide a survival shelter in case of a global catastrophe.
- Provide reliable space transportation systems to the Moon.
- Provide an isolated depository for high level wastes from the Earth in case of need.

[9] H.H. Koelle, *On the Past and Future of Lunar Development*, Report ILR Mitt. 299 (1996), http://vulcain.fb12.TU-berlin.de/ILR/personen/hh_koelle.html. The site contains many reports and other material on lunar exploration and exploitation.

The following secondary objectives are identified in the same paper:

- Improve the understanding and control of planet Earth.
- Stimulate the development of advanced technologies on Earth.
- Provide opportunities for international cooperation.
- Provide rewarding job opportunities.
- Assist in reducing tensions and conflicts on Earth.
- Provide the infrastructure and experience for global enterprises.
- Provide opportunities for involvement in frontier activities.
- Provide a peaceful output for the military–industrial complex.
- Contribute to the national prestige of the participating nations.
- Improve our understanding of the solar system.
- Improve our understanding of the Universe.

This list includes scientific, technological, economic, political and cultural objectives, some of which may be rather questionable, but nevertheless it constitutes a good starting point to assess the benefits of a lunar base for humankind as a whole.

To implement the base a 30-year programme is suggested. The base can be built and enlarged gradually; after that time the total crew on the Moon is expected to be about 120 people. The estimated cost is about US $100 billion over 30 years; it is only slightly more than the *Apollo* programme (US $70 billion in 1997 currency), but spread over a much longer period of time.

The planned annual expenditure is shown in Figure 5.9. Most of the expense is concentrated in the first 15 years, when the actual construction takes place. Figure 5.9 demonstrates that transportation costs are far higher than the value of what is actually taken to the Moon.

It is clear that any technological advances which reduce the cost of launching and then transferring the required payload to the Moon could be very effective in making the whole enterprise more affordable, while a reduction of the cost of the components of the base itself

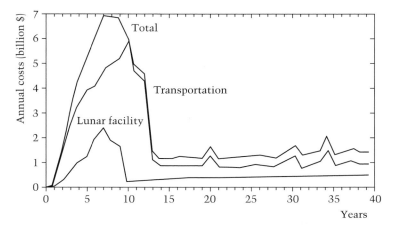

Figure 5.9. Estimated annual costs of building and maintaining a lunar base.

would seem to produce only marginal savings. For this reason, unconventional transportation technologies, such as those based on rotating tethers in Earth and Moon orbit and discussed earlier, have been suggested. If they prove to be feasible, the dream of building a lunar base will come closer to reality.

Cost considerations suggest that the construction of a lunar base will depend on the availability of heavy lift launchers, possibly expendable ones, of the 100 tonne class (to LEO) or even larger. The Energia rocket or a Shuttle derivative transportation system like that shown in Figure 2.7 are among the most realistic options.

A permanent lunar base has, with respect to a space habitat, several advantages. The gravitational field of the Moon, although considerably weaker than that of the Earth, makes life easier and avoids, at least partially, some of the health problems linked with weightlessness. Construction materials and some of the other consumables required, such as oxygen for breathing and as an oxidiser for rocket fuels, carbon and other elements needed by living beings to create a limited biosphere in a closed environment, and perhaps hydrogen as a fuel, can be made on the Moon and need not be transported from the Earth. A lunar base can be made radiation-proof using a layer of regolith, which is also a very good thermal insulating material.

An obvious advantage of a space habitat is the possibility of performing those scientific and industrial operations which, requiring microgravity, cannot be performed on the Moon. But an important thing from the psychological viewpoint can be the feeling of being on a planet, with a true up and down, and without the constant feeling of floating.

The frequently asserted psychological difficulty of getting used to a day–night cycle of almost a month is likely to be a minor problem only: the lunar base will have an artificial environment, in which a normal 24-hour day–night cycle could be maintained.

It is often stated that the very long lunar day has a drawback in that no known plant can survive in these conditions. If this were to be true, we would be forced to cultivate plants using artificial light, which is an inefficient and energy-intensive practice. By day, about 750 MW of light energy per square kilometre would be needed but, to obtain energy at night, either very long transmission lines or the extensive use of nuclear energy would be required. However, plants survive during the winter at high latitudes on Earth when only very limited light is available, and the day–night cycle changes considerably throughout the year. Even if artificial light is needed, plants could grow during the lunar day under sunlight with just a fraction of the 750 MW per square kilometre being required to keep them alive during the long night. Tests in a simulated lunar environment were performed by the Siberian Academy of Sciences, growing wheat under a month-long day–night cycle. They found that, if the temperature is lowered to slightly above freezing point, the plants 'hibernate' and resume their growth when the Sun reappears. The product was found to be similar to that under Earth conditions, since the plants grow continuously for the 14 days of sunlight.[10] Genetic engineering may make things simpler, by adapting some strains of plants to these peculiar conditions. Moreover, orbiting mirrors or greenhouses in the Moon's polar regions may be more practical than artificial lighting.

[10] R.C. Parkinson, *Citizens of the Sky*, 2100 Ltd., Stotfold, 1987, page 96.

From this viewpoint space habitats are better than lunar bases. Actually, there is some complementarity between lunar bases and space habitats. Some industrial operations will be performed in the latter, while the former will be needed to exploit the resources of our satellite. The dust which covers its surface is rich in many useful elements, such as titanium, which may be obtained without resorting to complex mining operations as on the Earth.

At a future date, it may be cheaper to extract raw materials from lunar dust and take them to where they are to be used, rather than to dig for them in deeper and deeper mines. But it is not only a problem of availability; on the Earth it is becoming less and less acceptable to use large areas of land for open cast mining or for industrial operations which have a large environmental impact. Moving these activities to the Moon would without doubt reduce environmental problems on Earth, yet without creating problems on a celestial body which naturally has extreme conditions.

The contribution of a permanent lunar base to the expansion of humankind into space will not only be that of a port or a shipyard, roles which can be better performed by orbital stations or space habitats, but mainly that of a logistic facility. An outpost such as that shown in Figure 5.10 will certainly not have the aspect of transparent domes connected by covered corridors which science fiction has made popular, for example the lunar city of Figure 5.2. But it will allow the start of scientific and commercial operations. The outpost will be complemented by pressurised vehicles for travelling around (Figure 5.11).

The expected products of the base will be first for its own use and then for export, mainly to space habitats but perhaps also to the Earth. They may include gases, such as oxygen and hydrogen, raw materials, construction materials (Figure 5.12), helium 3 as a potential nuclear fuel, thermal and electrical power, metals, ceramics, electrical materials and pharmaceuticals.

In time helium 3 may become a precious commodity. When humankind masters controlled thermonuclear fusion, probably using the deuterium–tritium reaction, the walls of the reactor vessel will be

Figure 5.10. An artist's impression of a lunar base (NASA image).

Figure 5.11. A pressurised rover for scientific experiments and for surveying for natural resources (NASA image).

Figure 5.12. An artist's impression of a mining base on the Moon, showing robots mining and processing regolith (NASA image).

strongly radioactive; they will have to be replaced often and then disposed of. However, the helium 3–deuterium fusion reaction is almost 'clean', but it is more difficult to ignite. Because there is no helium 3 available on the Earth, its presence on the Moon may be the *raison d'être* for importing goods to the Earth from another celestial body.

Helium 3 is rather rare on the Moon, about one part in 2,500 of the already scarce helium implanted by the solar wind in the regolith. To obtain 1 kg of helium 3 about 250,000 tonnes of soil must be processed. Based on the fact that a single kilogram of helium 3 would yield, with 60% efficiency, 100 GWh of energy, it is likely that, once suitable technology has been developed, helium 3 will be worth something like US $1 million per kilogram. At this price, the cost of bringing it to the Earth is not such a major consideration!

Perhaps even more important may be services including new scientific knowledge, engineering know-how, launch services, maintenance and repair of space transportation systems, waste storage services, training for space projects, tele-education and tele-entertainment, health care for particular types of diseases, and space observations for the protection of planet Earth.

With its activities growing, what started as an outpost and then became a permanent base could turn into a true colony, itself able to start new colonies in neighbouring regions. Among the challenges that lunar colonists will face is transport around the Moon. The Moon's low surface gravity makes the maximum speed a wheeled rover may achieve about 40 km/h. The absence of an atmosphere makes flying aircraft impossible, whereas ballistic rockets use much fuel, particularly as they must also use their engine for landing. Unconventional machines, like walking or hopping machines, may be the answer, but either their comfort or their speed is limited. Theoretically, the best solution is a magnetic levitation (maglev) train, a very expensive option best for a high density of traffic, which will only occur in the distant future, if ever.

Besides the scientific and technological issues, and among them medical and psychological issues, the economic, organisational and legal aspects must be given due attention. To succeed in an enterprise of such moment, incentives to attract private investors must be offered. The presence of private capital is needed both to add to the public funds and to ensure a continuity of funding which the latter sometimes lacks. A colony on the Moon would need help and economic assistance from the Earth for a long time, before becoming self-supporting, but this period of time could be reduced if governments and international organisations put policies in place to encourage its growth.

The surface of the Moon is an environment which is completely hostile to human life. People need to be protected, when outside the shelters, by space suits. On the Earth too, humans have settled in environments where life is impossible without protection; they have developed new technologies without which life would be impossible. In the context of the general technological level of their civilisation, the garments in which Inuits survive in their cold environment require no lesser technological feats than those, in the context of modern civilisation, needed to build a space suit for an astronaut in the lunar environment.

PRIVATE LUNAR BASES?

The lunar outpost discussed in the previous section is meant to be a government-sponsored enterprise or, better, an international enterprise in which the various governmental space agencies each play a major role. Large private industries would be encouraged through public–private partnerships. There are, however, some small private companies which have been founded with the aim of building and operating various space enterprises and, in this case, lunar settlements. The underlying rationale of such enterprises is that the political circumstances and ethos which led to the space race and, in particular, to the *Apollo* programme in the 1960s were exceptional. They are unlikely to be repeated in the future. The budgets of the space agencies are now shrinking. It may be a vain hope that in the future governments will allocate extra resources to activities in space. As the exploitation of space resources could become profitable, adventurous private investors are entering a new, high-risk arena – the exploitation of extraterrestrial resources. There is no doubt that this reasoning is based on many historical analogies, as almost all the exploration journeys of the past were motivated by possible commercial gain and were, either directly or indirectly, financed by private investors.

For example, the aims of the Artemis Project, a privately financed commercial venture to establish a permanent, self-supporting manned lunar base, are to ' ... develop lunar resources for profit, to demonstrate that manned space flight is within the reach of private enterprise and to create an environment for the growth of private industry in space'.[11]

Among the industries which such companies are promoting, space tourism is prominent. It is possible to speak seriously of tourism

---

[11] G. Bennet, *Artemis Data Book*, http://www.asi.org.

on the Moon (other than the telepresence mentioned earlier, which can be started right now) only after a base has been established to process lunar material and produce liquid oxygen and liquid hydrogen to be used as propellants. An order of magnitude reduction of the launch costs from the Earth is also needed. Moon tourism could become feasible when all these conditions have been met and a trip to the Moon and back can be sold for something like US $100,000.

There are legal aspects which affect the exploitation of the Moon, particularly for privately owned companies. The *Treaty on Principles Governing the Activities of States in the Exploration and Use of Outer Space, including the Moon and Other Celestial Bodies,* of 1967, ratified by 92 nations, states only general principles. An example is that the exploration and usage of outer space must be performed for the benefit of all countries without discrimination, and that states are required to encourage and facilitate international co-operation. Military operations and weapons of mass destruction are forbidden on the Moon, but military personnel performing peaceful duties are allowed. Claims of sovereignty by any nation are forbidden. This treaty was then followed by specialised agreements dealing with topics such as liability for damage, assistance for astronauts and many others.

A further treaty, *The Agreement Governing the Activities of States on the Moon and Other Celestial Bodies,* proposed in 1979, has been signed by only eight nations (Philippines, Uruguay, Chile, The Netherlands, Austria, Pakistan, Australia and Mexico), and not by the actual space-faring countries. The points which prevented its larger acceptance were the declaration that resources found on extraterrestrial bodies are a 'common heritage of mankind' and the proposal for an international regime to phase-in the benefits derived from those resources.[12] Perhaps some of the profits made by private organisations

[12] For a discussion concerning the fears of some developing nations that the exploitation of space resources could increase the gap between wealthy and poor nations, see W.K. Hartman, The resource base in our solar system, in B.R. Finney and E.M. Jones (editors), *Interstellar Migration and the Human Experience,* University of California Press, Berkeley, California, 1985. Little has changed in almost 20 years here.

working in space could be paid as royalties into an international development fund rather than setting up an international body chartered to exploit extraterrestrial resources.

In the present situation, there are only general guidelines which lead to the development of international space laws and regulations. Mainly dealing with safety, these should not hamper the exploitation of space resources. The difficulty remains of enforcing any restrictive rule against those who are actually able to work on remote celestial bodies. That should prevent activities in space from being regulated too strictly.

LUNAR POWER STATIONS

One of the products which a lunar colony could export to the Earth is energy. Goods that are most easily exported over large distances are those with low mass for unit value and whose transportation is fast. Energy (transmitted through a microwave link) and information are thus the ideal goods in this context, as they have no mass and travel at the maximum possible speed, the speed of light.

The cost of the energy produced on the Moon by solar panels and then transmitted to the Earth has been worked out to be less than 1 cent per kWh, lower than that produced by fossil or nuclear fuels, if environmental costs are taken into account.[13] This estimate is, however, a controversial issue. Such low costs might be possible only in a future in which the colonisation of the Moon is well under way. Lunar power stations (LPS) might thus be competitors of the space power stations described in Chapter 3, but their supporters say that they will be more efficient, from the economic viewpoint. The basic advantages are the same; there is very low environmental impact and little interference with other uses of the land or with other human activities. A lunar power station would be simpler than a space power station, as the panels could be directly laid out on the Moon's surface;

[13] D.R. Criswell, *Lunar-based Commercial Power and World Economic Growth*, 48th International Astronautical Congress, Turin, October 1997.

they do not require a supporting structure or an attitude control system. The station can be built using lunar materials and there is no need to construct electromagnetic launchers or their power systems. Moreover, the greater distance does not constitute a problem, as a microwave beam remains well focused in the space vacuum, although the receiving antenna on the Earth will have to be a large one.

The drawback will be that a lunar power station could produce energy only for half the time, as it would not work during the long lunar night. It would operate at peak power only at noon, if it is on the Moon's equator, i.e. when the Sun's rays fall perpendicularly on the solar panels. To produce the same quantity of energy as a panel in space, three equal panels on the lunar equator separated by 120° longitude from each other must be used. Power lines must connect the power stations to an antenna broadcasting the power to the Earth. Alternatively, it might be possible to put mirrors in orbit around the Moon, to reflect the microwave beams coming from the power stations which cannot broadcast directly. At the other end of the transmission line, on the Earth, a similar system of orbiting mirrors is needed to convey the microwave beams to the receiving stations when the Moon is below the horizon. The system is thus more complicated than that required for space power stations, but those who have studied such systems hold that it is more convenient, from the economic viewpoint.

6 Mars, the red planet

Martian exploration – the search for life on Mars – has grabbed the public's attention. And whatever results are found, they will astound us. But first, where on Mars is that essential ingredient for life, namely water? And is it safe to bring Martian material to the Earth for examination?

Should such a programme investigating Mars be a truly international programme? Should it use robots, or humans in space suits, or both? How could the best use be made of the resources available on Mars? And what is the optimum plan?

How long might it be before we can establish a human colony on Mars? What could such a settlement look like? Eventually, could we – or should we – change the environment of Mars to be more welcoming to the human species? If so, how?

DREAMS AND PROJECTS

The red planet has for centuries fostered dreams and legends. When astronomical observations started to unveil its mysteries, this trend could only be further strengthened. A hundred years ago it was a commonly held opinion that Mars, the planet next to the Earth away from the Sun, was home not only to living creatures but also to intelligent beings. They were considered to be more-or-less similar to ourselves.

A contributor, mostly unintentional, to these ideas was the Italian astronomer Giovanni Schiapparelli, who in the late nineteenth century observed through his telescope some thin dark features on the surface of the planet which he interpreted as straight lines. He described them with the Italian word 'canali', which was translated into

English as 'canals'. Canals refer only to artificial waterways, while the Italian word can be used for both artificial and natural water courses. This led to many speculations on the civilisation which might have undertaken such gigantic works of engineering, supposedly in an attempt to survive the process of desertification. While in his scientific papers Schiapparelli was very conservative and always adhered to the results of his observations, in some popular articles he unleashed his imagination, speaking of intelligent beings living on Mars, of their civilisation and even speculating on their political system.[1]

One person who contributed both to these fantasies and to scientific research on Mars was the American Percival Lowell, who became an astronomer, built a huge observatory in Arizona and wrote many books, eventually becoming the leading expert of his time – the beginning of the twentieth century – on the red planet. Due to his books, the idea that intelligent beings were living on Mars was generally accepted and this led to a good number of novels, from *The War of the Worlds* by H.G. Wells to *Under the Moons of Mars* by E.R. Burroughs, and from *The Martian Chronicles* by R. Bradbury to *Out of the Silent Planet* by C.S. Lewis.

Today, the images and data sent back to Earth by space probes leave no doubt. No intelligent beings roam the deserts of Mars, and it is likely that the planet does not host even the most primitive of lifeforms. Notwithstanding this, there are some who insisted that a rock formation, which in some pictures taken by a space probe looks like a human face, is actually a huge statue, a sort of Martian Mount Rushmore. For them the probe *Mars Observer*, which was launched from the Earth in September 1992 only to be lost in the vicinity of the planet, was destroyed by Martians.

Before images taken by a space probe showed Phobos (Figure 6.1), one of the two small (about 20 km in size) satellites of Mars, to be an irregularly shaped and cratered asteroid captured by the planet, some

---

[1] G. Schiapparelli, La vita sul pianeta Marte, extracted from *Natura ed Arte*, issue no. 11, year IV, 1895, see P. Tucci *et al.* (editors), *Giovanni Virgilio Schiapparelli, La vita sul pianeta Marte*, Mimesis , Milan, 1998.

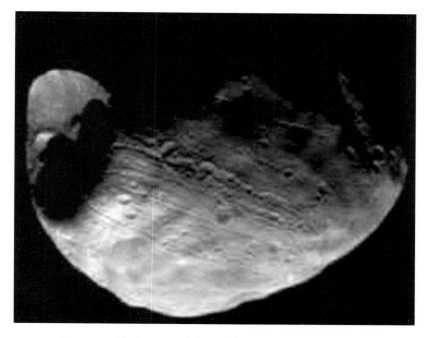

Figure 6.1. Phobos, one of the satellites of Mars (NASA photo).

scientists explained its low density by assuming that it was an artificial satellite, or, better still, a large space station built by the dy-
ing Martian civilisation as a sort of library or museum to preserve its
legacy. The composition of Phobos is very similar to that of a carbonaceous chondrite, a kind of very light meteorite; it may contain much
water, up to about 20% of its mass. As water can be decomposed into
hydrogen and oxygen, both of which can be used as propellant, it could
be a very useful 'refuelling' station on the way to or from the planet.

Apart from science fiction, the first serious and detailed study
for a manned space mission to Mars was described in the book *The
Mars Project*, by Wernher Von Braun, published in 1962. The suggestion was to launch a fleet of 10 spacecraft, with a total crew of 70, who
would stay on the surface of Mars for about 1 year. The same author
4 years later proposed a smaller project in which the number of spacecraft was reduced to two, one being a winged landing module, and the

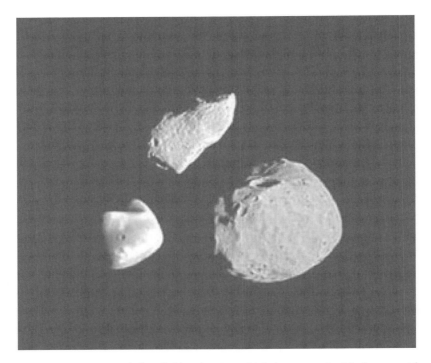

Figure 6.2. Phobos (left) and Deimos (right), compared with the asteroid Ida (top) (NASA composite photo).

crew cut to 12 people. Since then several projects have been proposed, sometimes together with statements from politicians supporting the missions and with some funding for basic research. In 1969, after the *Apollo* landing on the Moon, a project based on two nuclear propulsion spacecraft with a crew of 12 was developed. Once near the planet, eight people would land on the surface while four would remain in space, perhaps exploring the Martian moons Phobos and Deimos (Figure 6.2). The mission was supposed to leave Earth orbit on November 12, 1981.

The enthusiasm was short lived, and for years the hopes for an ambitious mission to Mars faded. Then in 1981, instead of setting course for the red planet as had been scheduled, a meeting open to all interested persons was organised to discuss possible scenarios for a mission to Mars. After the meeting an informal group, the 'Mars Underground', suggested that the manned exploration of Mars should be included in the goals of the American space programme. Thanks to the efforts of this group, the commission chaired by the astronaut Sally Ride included an expedition to Mars within the main goals of NASA, as noted in Chapter 1.

The US President George Bush, in a speech delivered on July 20, 1989, to mark the twentieth anniversary of the *Apollo 11* landing on the Moon, stated a new goal for the Nation: 'And next – for the new century – back to the Moon ... And then – a journey into tomorrow – a journey to another planet – a manned mission to Mars.' NASA launched the Space Exploration Initiative (SEI) and set up a committee which in 3 months prepared what was called 'The 90-Days Report'. This report stated that in order to fulfil the stated goal a 30-year programme had to be started. This included the expansion of the space station, then still only on paper, the construction of orbiting facilities to build large spacecraft, the construction of a lunar base for preparing propellants to be used on the journey to Mars, and only then a manned mission to Mars. The plan was evaluated and costed, and a figure of at least US $450 billion over 30 years came out. Even if this appears to be moderate compared with the military expenditure of US $800 billion per year over the whole world, it is so high that the whole project was thrown out; there have been no attempts to revive it.

The failure of the Space Exploration Initiative marked a turning point in the way in which a human mission to Mars is conceived. It demonstrated that an alternative to the large-mission approach, with huge spacecraft assembled in Earth orbit and large crews, had to be found. Such was the radical proposal put to NASA in 1990 by R.M. Zubrin and D.A. Baker and referred to by them as Mars Direct.[2] The name Mars Direct specifies the launch of the spacecraft from the

---

[2] R.M. Zubrin, D.A. Baker, Mars Direct, a proposal for the rapid exploration and colonisation of the Red Planet, in *Islands in the Sky*, Wiley, New York, 1996; R.M. Zubrin, Sending humans to Mars, in *Scientific American Presents the Future of Space Exploration*, vol. 10, spring 1999; R.M. Zubrin, *Entering Space: Creating a Spacefaring Civilization*, Tarcher/Putnam, New York, 2000.

Earth, without any in-orbit assembly, and the return flight from Mars without a rendezvous with a spacecraft orbiting that planet. This new concept was at least partially included in the Reference Mission devised by the NASA Mars Exploration Study Team in 1997.[3] This Reference Mission can be defined as a 'semi-direct' one, since the launch from the Earth is performed directly, without any in-orbit assembly. On the return journey the astronauts meet the interplanetary module in Mars orbit, in a way that is not very different from that used for the successful *Apollo* missions to the Moon.

The new approach does not rely on any synergies between the Moon and Mars space programmes, as was the case for the Space Exploration Initiative. This is good on the one hand, as a failure to obtain funding for one of the two enterprises does not affect the other, but, on the other hand, it does not exploit the potential savings which could result from a more integrated approach. At present, the Americans seem to be more interested in a manned mission to Mars, while the Europeans and Japanese are more for a 'first the Moon and then (perhaps) Mars' approach. This disparity could cause problems when detailed discussions to organise a joint international mission begin.

Even if some recent projects spoke of a lift-off in 2002 or in 2004, it is safe to predict that an expedition of humans to Mars will not be realised before 2020. What is first required is a clearly stated rationale for the programme, one which is accepted by the general public in the participating nations. They will then support it and contribute some taxes, looking forward to the day when they see, on their television screens, human beings walking on Mars.

THE 'MARS OUTPOSTS' APPROACH
Mars Outposts is not the name of a specific programme – rather it is an innovative approach to Mars exploration recently endorsed by the US-based Planetary Society. The core of this new philosophy is

[3] Information on the NASA *Reference Mission* can be found at
http://spaceflight.nasa.gov/mars/reference/hem/hem1.html.

an attempt to overcome the usual dichotomy between robotic and manned exploration and an effort to link the goals and requirements of single missions aiming for very ambitious results.

The single missions are, however, not very different from those already planned. They must be focused on a small number of sites, carefully chosen and then gradually equipped with telecommunications and navigation systems and, later, power and life-support systems. At these outposts there will be robotic rovers, balloons and exploration aircraft so that from them a detailed knowledge of the surface of the planet may begin to be gathered.

Another aspect of the Mars Outposts approach will be the participation of the public in the exploration of the red planet. The slogan of the whole enterprise might well be *we can all go to Mars*, not in person but using all the means which information technology now puts at our disposal. The images and the results of the exploration would be spread around the world through the Internet. People, particularly students and youngsters, would experience the feeling of exploring a new world. Such participation by the public in the exploration is the only way to prevent the Martian adventure from stopping as exploration of the Moon did in the early 1970s.

Among the early steps of the exploration programme, a sample return mission is needed. It has two goals – to study the composition of the soil and to search for microscopic life. To retrieve samples from Mars is a difficult task and, although missions of this type have been scheduled for the launch window of 2005, they may have to be postponed to 2009 or even later. Planning for a mission scheduled for 2005 was as follows:

1. Launch a lander, which deploys a robotic rover to search for some small samples. The rover brings the samples back to the lander, which transfers them into the nose cone of a small rocket – perhaps the size of an air-to-air missile – and puts them into orbit around Mars. The capsule containing the samples remains in orbit around the red planet.

2.   Two years later, in the next launch window, a new robotic lander repeats the same operation, resulting in a second capsule with samples in a Mars orbit.

3.   Towards the end of the same launch window as the second lander, another spacecraft is launched. This time it is an orbiter, which enters Mars orbit and captures the two capsules with the samples. The capsules are put into a larger re-entry capsule on top of a rocket, which will inject them into a Mars–Earth transfer orbit. The specimens will re-enter the Earth's atmosphere and land with an *Apollo*-style splash-down.

The mission is extremely complex, particularly the capture of the specimens in Mars orbit. Apart from the technical difficulties, there is another problem, namely the possible biological contamination of the Earth by Mars samples. One of the two possible extreme positions is that held by Robert Zubrin, and espoused in his various books.[4] Zubrin holds that there is absolutely no danger, because any parasitic organism has evolved to infect a particular type of life form. To use his words: 'humans do not catch Dutch elm disease and trees do not catch colds'. Moreover, a Martian living being, if such a thing exists, could not compete, on the Earth, with beings that have adapted to the terrestrial environment for a million years. Besides, Zubrin says, if Martian organisms exist, they already have been transported to the Earth many times in the past by meteorites, without problems. He concludes that those people who are afraid of Mars contamination would do better to leave the Earth immediately!

The opposite position is adopted by the ICAMSR (International Committee Against Mars Sample Return), an association that, although not opposing sample return missions to Mars in principle, suggests that the samples are not brought directly to the Earth but, instead, are left for a long quarantine period on the *International Space Station*. There they can be studied and, perhaps, stored for ever. The

[4] R. Zubrin, R. Wagner, *The Case for Mars*, Touchstone, New York, 1997, and R. Zubrin, *Entering Space: Creating a Spacefaring Civilization*, Tarcher/Putnam, New York, 2000.

committee holds that Mars is surely inhabited by micro-organisms which, like all bacteria, are potentially dangerous for any form of life. It believes that exchange of biological material between planets, even if that happens naturally via meteorites, is extremely dangerous. The committee considers that many epidemic diseases are caused by meteorites, comets and asteroids, and adopts a position similar to the strong panspermia idea introduced by Fred Hoyle and Chandra Wickramasinghe in the 1970s.[5]

Between these two extreme positions, most scientists and NASA's administration think that reasonable quarantine measures for all specimens originating from Mars must be taken. Even if Zubrin's arguments seem to be sound, the possible danger of Martian bacteria cannot be excluded and it is wise to take precautionary measures even if the risk is very low. The outside of the re-entering capsule containing the specimens reaches very high temperatures. That guarantees its sterilization, and it is not too difficult to design it in such a way that it arrives intact on the Earth. There are many biological laboratories on Earth able to open the capsule and study possible micro-organisms under the required safety conditions; there is a great deal of experience of viruses and pathogens of every kind, including those artificially created for bacteriological warfare. It could be much more dangerous to study the specimens in a space station, due to the difficulty of creating a sterile laboratory there.

The suggestion of sterilising the samples as soon as they reach the Earth is not viable, because one of the main goals of a sample return mission is to search for microbial life. To discover that would be the greatest scientific achievement of the mission – of all space missions to date. To kill the bacteria before studying them makes no sense!

A sample return mission may not solve the enigma of life on Mars:[6] we already have samples from Mars – the SNC meteorites – and

---

[5] The ideas of Fred Hoyle and Chandra Wickramasinghe are well described in the site www.panspermia.org.

[6] C.P. McKay, Life on Mars, Chapter 18 in A. Brack (editor), *The Molecular Origins of Life: Assembling Pieces of the Puzzle*, Cambridge University Press, Cambridge, 1998.

no conclusive evidence can be drawn from them. It is likely either that the specimens will contain nothing of biological origin, and so no evidence for life on Mars is found, or that they will contain some clues, which scientists could debate for many years. Only a scientific expedition to Mars and the construction of an outpost where biologists may work for a long time will give the final answer . . . keeping in mind that, if they find nothing, there will always be others who think that Martian life is just around the corner from the place where the samples were taken.

The first explorers on Mars will have to learn to live 'off the land', for this is the only way to make a manned mission affordable. The robotic outposts will be the ideal place for conducting experiments yielding the necessary know-how in this direction. It is there that the first experiments to produce the methane/liquid oxygen rocket propellant from some hydrogen carried from the Earth (or, better still, obtained from water found on Mars) and carbon dioxide from the Mars atmosphere. This propellant would be the main fuel for all the activities of the first humans on Mars, and for the rocket by which they may return home.

NASA plans to send a lander (Figure 6.3) with all that is required to grow some plants using Martian soil and, possibly, Martian water. Some plants may be genetically engineered, by introducing into them some genes from a jellyfish so that they will glow a faint green colour if they encounter problems. A highly symbolic experiment would be to grow a flowering plant from seed using Martian soil and carbon dioxide and water from the planet's atmosphere.

The camera aboard the lander will send back images of the plant, so that – if it does not wilt – its growth may be measured; for our progress, it needs to thrive in the Martian environment. It is worth pointing out that a lander with a plant does not violate the non-contamination treaty signed in 1967 by the United States and the (then) Soviet Union. That stated, among other things, that no contamination of the Mars environment from terrestrial biota should be

Figure 6.3. A mission to Mars suggested for the 2005 or 2007 launch window (NASA image).

allowed for a period of 20 years. It was believed that 20 years would be enough to determine whether Mars had any forms of life and, if so, to put an adequate protection policy in place.

A contamination probability of less than one in a thousand is considered a sufficient precaution. Landers are to be scrubbed with alcohol and the assembly performed in a cleanroom, taking surgical-type precautions. The spacecraft would also be heated, to sterilise it by a factor of about 10,000. Particularly in the light of the strongly oxidising nature of the soil, as shown by the *Viking* experiments, this procedure is considered sufficient.

A hint of the difficulty of sterilising space probes came when the *Apollo 12* astronauts, Pete Conrad and Alan Bean, brought back to the Earth the television camera of the *Surveyor 3* probe that had landed on the Moon almost 3 years earlier. Analysis revealed that a colony of *Streptococcus mitis*, which evidently survived all the pre-launch sterilisation procedures, was still alive. These bacteria had withstood

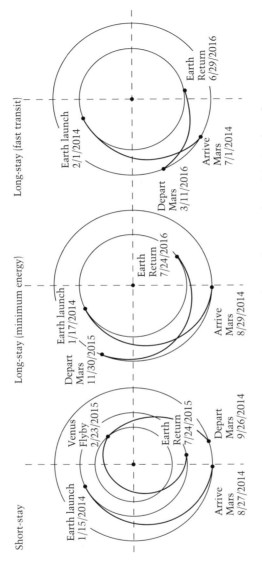

Figure 6.4. Sketch of the trajectories, with the Sun at the centre, for one short-stay (opposition) and two long-stay (conjunction) Mars missions. The difference between the latter two is linked with the propulsion requirements, the fast-transit mission requiring a greater velocity than the minimum-energy mission. The dates are given using the US calendar convention of month/day/year.

the lunar vacuum and harsh radiation environment and huge temperature excursions for 31 months. However, no contamination of the outside of the spacecraft or of the lunar environment was found.

Experiments in which plants are grown on the planet Mars can be very useful to avoid fears of Earth contamination by Martian samples. For if there are dangerous lifeforms on Mars, their effects on the plant should be evident.

Later on, larger power generators will be installed at the robotic outposts, and the production of propellant will start. A large greenhouse could encourage an ecosystem able to support life for the first astronauts. Then the time will be ripe for the first human beings to land on the red planet.

MISSION PLANNING

The alternatives for a manned mission to Mars are either a short mission, with a stay of less than one month, often referred to as an opposition mission, or a longer one, with a sojourn of about a year or even more, a conjunction mission.[7] The spacecraft trajectories for one short-stay and two long-stay missions to Mars, referred to the launch window of 2014, are shown in Figure 6.4. The time spent in space and on the planet for these three missions is given in Table 6.1.

The first mission profile is typical of the proposals made from the 1960s to the 1980s. In an attempt to minimise the total mission time, a good launch window for the return journey is not waited for and the inbound leg, via the planet Venus, is quite long. Alternatively, a long outward journey, with a flyby of Venus, may be chosen, while the shorter route may be used for coming back home. Whilst the total time is the shortest possible for a minimum energy mission, the time on the Martian surface, i.e. the useful time for accomplishing the mission's goals, is very short. The time spent in space, which can only be used for making medical observations, is long. Moreover, the danger to the crew – apart from the launch – comes mostly from the

[7] Opposition indicates that Mars and the Sun are on opposite sides of the Earth, whereas conjunction indicates that they are on the same side.

Table 6.1. *Time (in days) spent in space and on the surface of Mars for the three missions considered in Figure 6.4*

	Short-stay (opposition)	Long-stay (conjunction) (minimum energy)	Long-stay (conjunction) (fast transit)
Outbound	224	224	150
Stay	30	458	619
Inbound	291	237	110
Total	545	919	879
Total in space	515	461	260

time spent in space, owing to their exposure to radiation and to the long periods of weightlessness. To help the crew to remain in good health, they will have to follow a severe exercise regime. Further, it is likely that some form of artificial gravitation will be required, and this greatly increases the complexity and the mass of the spacecraft. The dangers due to radiation are further increased in the inner solar system approaching the orbit of Venus for the flyby. The need for a thicker shield against solar energetic charged particles when travelling so close to the Sun adds further to the mass of the spacecraft.

The second mission profile leads to a far longer mission, but the time available for doing useful work on Mars is multiplied by a factor larger than ten, while the time spent in space is reduced by roughly 10%. The spacecraft never gets closer to the Sun than the Earth's orbit. By increasing slightly the expenditure of rocket fuel energy at both injection points, into the trans-Mars and trans-Earth trajectories, the substantial reductions of transit times shown in the last column of Table 6.1 can be obtained, although they must be accompanied by an increase in the time on the surface of Mars waiting for a suitable launch window. For the third mission profile, Table 6.1 shows that the transit time becomes almost half that of the short-stay mission, while increasing the useful time on the surface to more than 600 days.

A completely different mission profile – a controversial one – would be for a one-way mission to Mars, with the immediate objective of humans settling there and colonising the planet. Such a mission would have to meet many difficult requirements, but the trajectory and the fuel to return to Earth would not feature amongst them.

The choice of the type of mission influences many issues. For example, a short-duration mission has the possibility of aborting during the outbound leg. The spacecraft could reach as far as Mars and then inject directly into a return trajectory without landing, just as *Apollo 13* did, passing behind the Moon and coming back. No long-stay mission can do that, and the concept of 'abort on the surface' has to be developed. If something goes wrong during the outbound journey the crew will have to land on Mars, and wait there for an opportunity to come back or for a rescue mission to be sent. This is actually fairly safe. It is better to wait on the surface of the red planet than be in a crippled spacecraft on a long trajectory through the solar system, particularly if a previous unmanned mission has put a habitat on the surface of Mars, with all the provisions needed for survival. Long-stay missions are made safe by landing a shelter and having a return vehicle ready on Mars before humans are dispatched there. Another strategy would be to plan for a long-stay mission and, if something goes wrong, to abort the mission as a short-stay one; the outbound trajectories for the two types of missions are not so different from each other.

In a classical short-stay mission one or more vehicles with part of the crew remain in orbit around Mars, as in the *Apollo* lunar missions. This is not feasible for long-stay missions, because the orbiting crew would spend more than 1 year in space, soaking up excess radiation without having much that is useful to do. Thus the mission must be either direct, with the return journey directly from the surface, or semi-direct, with a rendezvous in Mars orbit with an unattended spacecraft. This does not pose a problem, with current automated spacecraft technology.

A further important point is that the crew might exploit some local resources, or, as is usually said, 'live off the land'. Some of the supplies which the crew will need on Mars need not be carried from home, even the propellant for the return journey. In the Mars Direct project, for instance, a launch vehicle (without propellant) for the return flight, a 100 kW nuclear reactor, a small chemical plant, two small automatic vehicles and 6 tonnes of liquid hydrogen are launched 2 years before the actual manned mission starts. As soon as it arrives on Mars, the nuclear reactor is unloaded automatically and located at a suitable site. The compressor of the chemical plant then starts compressing carbon dioxide, the most abundant (95%) component of the Martian atmosphere, and storing it. From the carbon dioxide and the liquid hydrogen carried from Earth 29 tonnes of methane plus some water are obtained; the water is transformed into oxygen and hydrogen by electrolysis using electric power produced by the reactor. No cryogenic fluid (liquid hydrogen) needs to be stored any more. The hydrogen next reacts with carbon dioxide, and more carbon dioxide is decomposed into oxygen and carbon monoxide. The latter can be discarded and the final product is about 108 tonnes of oxygen–methane propellant. The whole operation would be concluded in 18 months from the launch date, 8 months being for the flight to Mars and 10 for the production of the propellant. The mission planners still have some months to check that everything is all right before the first astronauts and cosmonauts begin their journey to the red planet.

To take a return vehicle and a habitat to the surface of Mars and then to land the crew 2 years later exactly at the same spot, automatic landing techniques have to be mastered with exceptional precision. This is, however, not impossible. As shown in Figure 4.1, the *Apollo 12* Lunar Excursion Module landed close enough to the *Surveyor 3* probe for that to be visited, and that was with technologies which are now more than 30 years old. The aerobraking manoeuvre performed successfully by the *Mars Pathfinder* lander in 1997, although not by the *Mars Climate Orbiter* in 1999, required a similar precision. By the time the manned expedition to Mars is launched, the experience gained with other robotic space probes should make such precision landings almost routine.

The Mars Direct project takes into account the possibility of several accidents, such as a landing in a place far from the desired destination. A vehicle propelled by an internal combustion engine working with methane and oxygen, able to travel for about 1000 km, would be transported together with the crew. This could reach the return vehicle by travelling across the Martian surface. Moreover, a second automatic spacecraft with a return vehicle would be launched at the same time as the manned vehicle, bound for another point on the surface of Mars in preparation for another expedition 2 years later. In an emergency it could be redirected to the place where the crew had landed and then start immediately to prepare the fuel needed for the return journey. If everything turned out disastrously, the crew would remain in their habitat for the 2 years needed to organise and mount a rescue expedition.

THE FIRST HUMAN BEINGS ON MARS

If an expedition to Mars is performed within the first few decades of the third millennium, it is very likely that it will be based on the NASA Reference Mission described in the previous section.

As this will not include the building of spacecraft in Earth orbit, a giant rocket, termed a heavy-lift launch vehicle, will be required. The promoters of the Mars Direct project proposed a large chemical rocket, named Ares, capable of launching 47 tonnes towards Mars. As this rocket could also be used as a non-reusable launcher for other missions, cost savings linked to multiple production would be obtained. Such a rocket would be able to launch 121 tonnes into low Earth orbit, slightly more than the almost 100 tonnes of the current Russian rocket Energia, to which the Ares bears some similarity, and the 108 tonnes for the Saturn V which launched the *Apollo* missions. The total mass at launch would be of 2,200 tonnes, less than the 2,400 tonnes of Energia or the 2,800 tonnes of Saturn V. The NASA team considered many options, including a modified Energia rocket, a Space Shuttle derivative (Figure 2.7) or even a rocket derived from the now almost forgotten Saturn V, with due modifications. The team favoured the second suggestion as the most convenient and practical option.

The mission would start with the launch of one or two unmanned spacecraft to deliver provisions for the crew, a fuelled return vehicle and a habitat to the red planet. If the option of sending an empty return vehicle and manufacturing the propellant on Mars were to be followed, a single spacecraft might suffice.

So long as the phases described above had been carried out successfully, two more spacecraft would be launched in the following launch window. One would be an unmanned return vehicle while the second would carry a crew of four astronauts, the latter being launched about two weeks before the former.

The size of the crew is a critical issue. The ambitious short-duration mission of the Space Exploration Initiative would include more than 10 people, perhaps more than 20. Amongst the astronauts there would be scientists and at least one medical doctor. With today's approach there would be no more than four or five astronauts, each a scientist but trained in at least two disciplines. The highly automated spacecraft would not require specialised astronauts solely for guiding the vehicle; a medical doctor would probably not be included in the small team. Piloting and curing diseases would be performed as 'second jobs' by some mission specialists.

After reaching Earth orbit, the trans-Mars injection rocket stage would be fired. The NASA Reference Mission suggests the use of three nuclear engines, directly derived from the NERVA fission rockets of the 1960s (Figure 4.6). This allows a somewhat reduced transit time, and increases the payload sent to Mars. The more conservative Mars Direct mission advocates the use of chemical rockets to avoid the delays associated with the construction of innovative hardware and the potential political or environmental problems of a nuclear-powered rocket. There is no doubt that, in time, it would be really worthwhile to exploit the high performance of nuclear propulsion. Even if not used for the first journey to Mars, it could be very useful for the later exploration of Mars. Studies on thermal nuclear propulsion are now slowly growing again, and there are hopes that a nuclear rocket will be available in time for the first crewed mission to Mars.

If the travel time does not exceed a few months, the astronauts can manage without artificial gravity, as experience on the *Mir* space station has clearly demonstrated. Otherwise, artificial gravity approaching (or equal) to that on the surface of Mars, i.e. three eighths of that at the Earth's surface, could be produced by spinning the spacecraft, possibly made of two sections connected by a long tether. Because the crew will live in the same habitat as they will use on the surface of the planet, no additional living quarters would be needed (Figure 6.5).

To go into an orbit around Mars is a tricky manoeuvre which will be performed by aerobraking in the atmosphere of Mars (Figure 6.6). Precision aerobraking is already a tried and tested technology.

Figure 6.5. Artist's impression of the habitat landing on Mars. The two-storey cylinder has a diameter of 7.5 m, giving ample pressurised space for the crew to live in. The wheels move the habitat to join it to another similar one to create a larger pressurised dwelling (NASA image).

Figure 6.6. Artist's impression of a spacecraft aerobraking in the atmosphere of Mars, as seen from the Martian surface (NASA image).

Mars Global Surveyor even had its aerobraking procedure modified in such a way that the aerodynamic forces could be precisely adjusted to avoid stressing a damaged solar panel.

Once on the Martian surface, the habitat which carried the crew through interplanetary space and that which was transported to Mars 2 years earlier (which could act as an emergency shelter in case the former was badly damaged during landing) could be joined, to form a larger dwelling, with more than 200 square meters of habitable floor area (Figure 6.7).

The habitat has to be pressurised because the atmospheric pressure at the Martian surface is only 0.6% of that at the Earth's surface. The comfort of the crew could be increased by an inflatable structure having more available space, for example such as that shown in Figure 3.6. The scientific duties of the team to be landed on Mars would in all probability include short- and long-range reconnaissance

Figure 6.7. Artist's impression of the two Martian habitats ready to be joined together (NASA image).

missions, using pressurised 'rovers' for planetary research of many types. They would include experiments on the use of Martian resources to produce oxygen and methane to be used as fuel. The search for sub-surface water and attempts to grow plants using Martian soil (and possibly water) will be other important challenges. One of the most important goals is, however, that of finding convincing evidence for the presence of Martian lifeforms, or for their absence. The search for at least fossil life will involve looking underground, in the permafrost layer which is expected to exist. That could require long journeys from the outpost, located in a zone flat enough to make the landing of spacecraft not too risky. One such journey would be to the floor of one of the deepest canyons, like Marineris Vallis, where it is more probable that liquid water may still exist as moisture, at the surface or immediately below it, at least at certain times of the year, than elsewhere.

The anticipated stay would be of 1 year and 4 months, after which the return voyage of almost 8 months would start. The

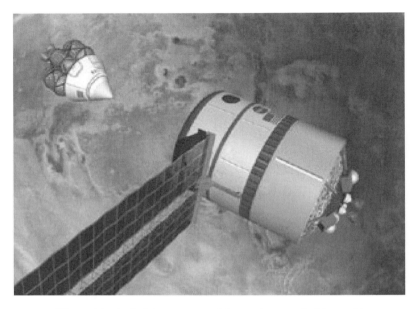

Figure 6.8. Artist's impression of the rendezvous, in Mars orbit, between the return capsule and the habitat which will be home for the astronauts on their journey Earthwards (NASA image).

reference mission includes the launch of a return capsule from the surface and the rendezvous in Mars orbit with a return habitat (Figure 6.8), ready to be injected into the trans-Earth trajectory. The return flight will end with a direct plunge into the Earth's atmosphere, where the manned capsule will be slowed down by air drag and then by parachutes, to land in the same way as the *Apollo* capsules did.

The Mars Direct project was conceived with a tight budget in mind, with only one habitat on the surface (Figure 6.9). For the return journey the crew is packed into the Mars ascent vehicle, to avoid the need for a rendezvous with an orbiting, larger habitat.

The cost for the NASA Reference Mission (1997) was estimated to be about US $50 billion, while Mars Direct was evaluated at US $30 billion. Zubrin[8] proposed a lower estimate, down to US $6 billion, if the whole mission were carried out by private enterprise, operating

[8] R. Zubrin, *The Case for Mars*, Touchstone, New York, 1996.

Figure 6.9. Base for the first Mars Direct mission, from a painting by Robert Murray. Note the two-storey habitat obtained from the landing module, the return vehicle, the nuclear reactor in the crater in the background, an inflatable greenhouse and a wheeled vehicle (S. Schmidt and R. Zubrin *Islands in the Sky*, 1996. This material is used by permission of John Wiley & Sons, Inc.)

with the procedures of successful commercial companies and not of government agencies.

The approach described here involves only established technologies and launchers of the same class as existing ones. If it must be criticised, it could be said that it is too closely linked to the use of conventional technology, but this is a sign of the times. To land a man on the Moon in 1969 there were enough resources to perform much research and development work, while today for the Martian adventure we must make do with what is already available. Otherwise, the mission will be postponed again and again.

THE BEGINNING OF COLONISATION

After the experience of the Apollo missions, one of the mistakes which must be avoided is that of undertaking a number of missions to Mars, each one being seen as a reconnaissance on its own, without planning for a permanent human presence on that planet. One of the real advantages of the Mars Outposts approach and that of either the Reference Mission or the Mars Direct project is that they include a number of coordinated missions, with the aim of establishing several permanent bases on the surface of the planet.

After the return to Earth of the first manned mission, a second one should be planned for the following launch window, i.e. with a second mission leaving planet Earth two years after the first launch. The Reference Mission suggests that the second crew should land in the same place as the first one. The third habitat would then be connected to the first two, and the rovers and all other equipment could be refurbished and reused. As a result, after more than three missions a major outpost would be created on the red planet.

On the other hand, in the Mars Direct plan the second crew is expected to land about 300 km from the first, where 2 years earlier an automated cargo vehicle had landed and where now a fuelled return vehicle would be ready and waiting. That cargo vehicle would be ready for a future mission. Going on in this way, after a few years and at the cost of only two launches every 2 years, there would be a number of small outposts, at distances of 300 km from each other, which could be visited by other crews during their stays on the planet. The project anticipates that, after the third launch, the efforts would be intensified with the launch of a further Ares rocket every 2 years. At a certain point, the rocket would be upgraded with the addition of a third stage.

Within 24 years a permanent settlement with 48 people operating simultaneously on the planet would have been established. Powered by a 18 MW power station, the habitable space would be of more than 160 m^3 per person (Figure 6.10). An artist's impression of a busy Mars base, located at the foot of the volcanic Pavonis Mons is shown in Figure 6.11. At this point, true colonisation of the planet could start, building living spaces, industrial and agricultural plants, stores and other facilities. The main settlement would be surrounded by a

Figure 6.10. Sketch of a Mars Direct settlement, showing a number of habitats connected together to form the first nucleus of a colony (S. Schmidt and R. Zubrin *Islands in the Sky*, 1996. This material is used by permission of John Wiley & Sons, Inc.)

Figure 6.11. Artist's impression of a Mars base, located at the foot of the volcanic Pavonis Mons (NASA image).

number of small outposts, each being home to a few people for a short time during their travels over the surface.

An improved version of the Ares rocket, significantly referred to by the authors of the project as the *Martian Mayflower*, could at this point transport 12 colonists on a one-way journey. Colonisation, by its very nature, requires mainly one-way journeys or, better, a transportation capability on the return journey which is much smaller than that outward bound.

The scenario outlined above can actually lead to the settlement of a certain number of people on Mars and to the beginning of economic activities there, starting the long road towards a self-supporting community. All this can be done using current technologies and at a cost which is consistent with today's economic situation. A project of this nature will surely be carried out by a number of nations, but this is more important from the political than from the economic point of view. The costs are such that they are not out of range of the United States, the European Community, Japan and Russia, once the economic problems of the latter have been solved.

It has been said that one of the advantages of the Reference Mission, and of the new NASA approach to a manned Mars mission, is that of not requiring operations on the Moon's surface before starting out for Mars. However, manned missions and the construction of outposts on these two celestial bodies can be complementary; a strategy of the 'first the Moon, then Mars' type, like that generally considered in Europe, can lead to significant savings. One of the most costly aspects of both enterprises is the development of a new heavy-lift rocket. If the cost of that could be shared between the two programmes, it would be much more reasonable.

From the graph presented in Figure 5.8, it is clear that the cost of building and maintaining a lunar base drops sharply after about 15 years. Thus, it could be a wise strategy to start by returning to the Moon and building an outpost there and then a permanent base. When the costs start decreasing, and as the lunar colony becomes at least partially self-supporting, the Martian adventure should start. At this point lunar factories would be able to supply some of the required liquid oxygen fuel beyond the Earth's 'gravitational well', i.e. the region close to a large planet (in this case the Earth) from which escape is costly in energy terms. This would lead to further savings. Also, the experience gained on the Moon should make the construction of the Mars outpost much easier.

As already stated, no radically new technology is needed to land humans on Mars and to bring them back home safely. However, it would be advisable to seek an alternative technology enabling human missions to Mars with lower transit times, and reduced inconvenience and risks. The experience which will result from the assembly and the operation of the *International Space Station* will be invaluable when preparing for human missions to Mars. It could even become a servicing centre for interplanetary spaceships. Further, the availability of such a laboratory in space might allow us to study propulsion systems other than those using chemical fuels. The first of such new technologies is nuclear rockets, which have actually been considered in the Reference Mission. Systems based on nuclear fusion which can lead to a true revolution in solar system transportation are presently under study. An antiproton catalysed microfission/fusion propulsion system has been described in Chapter 4 (Figure 4.8); it would allow extremely short transit times to Mars. For example, for the 2005 launch window to Mars, a launch from Earth on August 12, 2005 would arrive at Mars on September 29, 2005, a journey of only 48 days! The return journey starting on October 29, 2005 would arrive back at Earth on December 10, 2005. The whole mission would last about 4 months, with a full month being spent on the surface of the planet.[9]

A much simpler, nuclear fission rocket fuelled by americium 242 (as studied by the Project 242 of the Italian Space Agency) and

---

[9] R.A. Lewis *et al.*, *Antiproton Catalysed Microfission/Fusion Propulsion Systems for Exploration of the Outer Solar System and Beyond*, First IAA Symposium on Realistic Near Term Advanced Space Missions, Torino, June 1996. See also Chapter 9.

producing a large thrust has the potential for a short transit time to Mars. There is no doubt that, in the long run, it would be simpler to establish a human settlement on Mars by being able to travel between the two planets at any time, rather than having to wait for the next launch window which recurs at intervals of every 2 years or so.

Other possibilities which can be investigated include electric propulsion, with the electric power being generated by solar panels or by nuclear reactors, and solar sails (Chapter 4). These are particularly well suited for automated space transports. Solar-electric cargo ships, operating continuously between the orbits of Earth and Mars, could greatly reduce transportation costs and increase the availability of useful materials on Mars. The transit times of such vehicles would be longer, of the order of 2 or 3 years, but the possibilities of carrying larger habitats, a larger number of vehicles and more fuel would make the stay of the first Martian explorers more comfortable and profitable.

In the more distant future, but on time scales not incompatible with missions to Mars, large solar sails might be built in Earth orbit; they could provide low-cost transportation toward the orbit of Mars.[10] Such sailing ships could continuously commute between the two planets so that a reasonable number of spaceships would provide a good Earth–Mars transportation service. Sail ships do not have the strict launch window limitations of chemical rockets using elliptical transfer orbits around the Sun, and so it is not necessary to wait for the customary period of slightly more than 2 years between one launch and the next.

It would, however, be a mistake to pin too many hopes on technological advances and to wait. If this logic had prevailed in the 1950s and all space activities were frozen waiting for the technological wonders which were then expected in the 1980s, not only would nobody have yet set foot on the Moon but we would have neither telecommunications nor weather satellites, which are of great benefit to human society. What today may look like science fiction could materialise in

[10] J.L. Wright, *Space Sailing*, Gordon and Breach Science Publishers, New York, 1994; C.R. McInnes, *Solar Sailing: Technology, Dynamics and Mission Applications*, Springer-Praxis, London, 1999.

a few years into high-tech hardware. While continuing with missions planned with today's technology, we must support advanced research in non-conventional techniques, especially for propulsion. In the future dramatic breakthroughs might occur.

A PLANET TO BE TERRAFORMED

The images taken by the *Viking*, *Pathfinder* and *Mars Global Surveyor* probes from orbit around Mars and on the surface of Mars are very impressive. Mars really is a land of contrasts. It has the largest volcano in the solar system, Olympus Mons, 25 km tall (but there is no active volcano on Mars now), and a canyon, Valles Marineris (Figure 6.12), which would really dwarf the Grand Canyon on the

Figure 6.12. The planet Mars, with the giant canyon Valles Marineris very clearly shown (NASA photo).

Earth. It bears the traces of impressive and dramatic events in the past, which have remodelled its surface – the northern lowlands, Vastitas Borealis, probably due to the impact of a large meteorite, the Tharsis Bulge, probably of volcanic origin, with three huge volcanoes (Pavonis, Arsia and Ascraeus Mons), and a huge number of impact craters, chasms, and mountains. The poles are covered by 'dry ice' caps (frozen carbon dioxide) which shrink in the summer and grow in the winter. Beneath, it is probable that there are large amounts of water ice mixed in the soil.

The difference between the two ice caps is due to the fact that the orbit of Mars around the Sun is much more elliptical than is the Earth's and the inclination of the planet's axis of rotation makes the seasons far more extreme in the southern hemisphere than in the North. This is also believed to be the explanation for the occurrence of violent dust storms which can last for months at a time.

Mars is smaller than the Earth, its surface area being about as large as the sum of all the continents of our planet. A simplified map of the planet is given in Figure 6.13. Note that the zero height

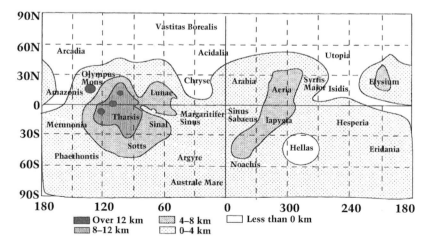

Figure 6.13. Simplified map of the Martian surface showing heights above an arbitrary level (R. Zubrin, *The Case for Mars*, Touchstone, New York, 1997).

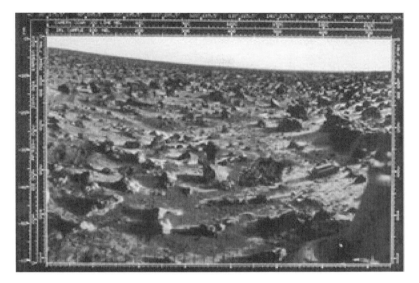

Figure 6.14. Utopia Planitia, the landing site of *Viking 2* (NASA photo).

level has been arbitrarily set, as there is no sea for a mean sea-level reference.

The regions in which the two *Viking* probes landed in the mid 1970s, Utopia Planitia (Figure 6.14) and Chryse Planitia (Figure 6.15) are beautiful, particularly at sunset when the colours become even warmer and the pink sky turns to a deep red. The beauty of the red planet, which seen from the surface is actual, must not deceive: even if it is less inhospitable than the Moon, the surface is made up of barren deserts, compared with which any desert on Earth looks like a pleasant place.

More than the red sand and the boulders scattered all around, which look like the volcanic stones of rocky deserts on the Earth, or the sky, pink due to the particles of sand rich in iron oxide carried aloft by the wind, its atmosphere is what makes Mars an alien planet for humans from the Earth. It cannot be breathed, as it is almost entirely carbon dioxide, and also it is much too thin. The pressure at the ground is less than one hundredth of the atmospheric pressure

Figure 6.15. The landing zone of *Viking 1* in Chryse Planitia (NASA photo).

on Earth[11] and has much variability with latitude. It offers very little protection from the Sun's ultraviolet radiation. And there is only very limited protection from cosmic rays due to the almost complete absence of a planetary magnetic field. From this point of view Mars is only a slightly better place for life than the Moon, even if the thin atmosphere scatters light and gives to the planet a pleasant aspect which is different from the sequence of black shadows and blinding light typical of conditions on the Moon. The temperature at the ground as recorded by *Viking* probes varies between $-14\,^{\circ}$C and $-120\,^{\circ}$C. Thus, liquid water cannot exist on the surface and most of the frost which the *Viking* landers saw is carbon dioxide clathrate (Figures 6.16, 6.17).

[11] The pressure at the ground on Mars is similar to that at an altitude of 35 km in the Earth's atmosphere, in the middle of the stratosphere.

Figure 6.16. Picture taken by the *Viking 2* lander showing frost in Utopia Planitia. The frost does not appear to be frozen carbon dioxide (dry ice) but is more likely to be a carbon dioxide clathrate (six parts of water to one part of carbon dioxide) (NASA photo).

Figure 6.17. Late-winter frost on the ground of Mars around the *Viking 2* lander in Utopia Planitia (NASA photo).

Water ice is, however, present in the polar caps – together with large quantities of frozen carbon dioxide (dry ice).

Humans moving around on the surface of Mars must wear suits similar to the familiar space suits of astronauts; they must carry with them the oxygen which they need to breathe. However, the low gravity value means that they can carry this equipment without being too hampered by it, and the lunar experience has shown that humans can adjust surprisingly well to this type of environment. Human beings will lope over the barren Martian landscape as they did over the Moon. As mentioned earlier the habitat and the rover vehicles must be pressurised; to enter them one has to pass through an airlock.

If colonisation is to proceed beyond a scientific station, or a little more than that, it will be necessary to consider the idea of terraforming the planet. The term *terraforming* introduced by Isaac Asimov has been widely used by science fiction writers to designate the transformation of the physical and environmental characteristics of the surface of a planet to make it suitable for supporting human life. The very idea of such an operation has such huge implications that it seems bound to remain in science fiction novels more than in serious scientific/technological studies, but actually this is not the case. It has recently been realised that to terraform a planet may be easier

Figure 6.18. Depiction of an inflatable greenhouse on Mars (NASA image).

than had been previously thought. And so the word terraforming has entered our technical vocabulary, like many other words invented by science fiction writers, *robot* being one of the better known.

Even before starting the actual terraforming process, we could try to seed the planet's surface with some lifeforms which can endure the very harsh Martian conditions. After the first human landing (or even before, if the Mars Outposts strategy is adopted), the first attempts at growing plants on Mars will be made. Normal terrestrial plants might thrive on the Martian soil in a pressurized greenhouse (Figure 6.18). The soil is regolith, just dust from rocks on the ground, with absolutely no organic matter. It must be enriched with fertilizer and with the organic substances which plants need. The soil may contain oxidising agents, harmful to plants, which would first have to be removed. If water obtained by melting Martian ice is used, that might have to be purified. Genetic engineering might be used to adapt plants to the Martian conditions, allowing a lower pressure in the greenhouse or reducing the need for modifying the soil. But, sooner or later, an attempt to farm the Martian soil outside greenhouses will have to be made. That raises ethical issues which will be dealt with later.

Some primitive terrestrial bacteria can survive in extreme environmental conditions, but it is unknown whether they could do so on Mars. In particular, they may not withstand the very low night-time temperatures, the radiation not stopped by a magnetosphere and the ultraviolet light not stopped by an ozone layer. Research work to identify the best bacteria has already started at the NASA Ames Research Center: if no suitable lifeform is found, the only option will be genetic engineering. There is no doubt that the terraforming process could be greatly speeded up if the conversion of Martian regolith into soil could be started before the environment changed much.

Some bacteria (like *Micrococcus radiodurans*), able to survive very strong radiation doses, 10,000 times the dose which a human being can withstand, have been discovered inside nuclear plants on Earth. By genetically engineering these it may be possible to obtain many useful products, such as medicines, so precious for the first Martian colonists.

Owing to the characteristics of the planet, the work of terraforming can be divided into two phases – increasing the atmospheric pressure, perhaps by heating the surface, and making the air breathable. The first step seems to be easier than the second, as there is evidence that in the very distant past the atmosphere of Mars was far thicker than it is now, even to the point that the pressure at the surface was twice the atmospheric pressure on Earth, and the planet was warmer. The first phase would then be a sort of planetary restoration, aimed at giving back to Mars the atmosphere which it had some 4 billion years ago.

The way to do this is to use the *greenhouse effect*, i.e. the capture of some of the heat radiated from the planet into space, and which leads to an increase of the temperature at its surface. We know well, maybe even too well, that the increasing greenhouse effect is an actual danger for the Earth, and that some molecular gases, not harmful to humans, are very effective here. Carbon dioxide and water vapour are two important greenhouse gases, as are methane and ammonia. Even though the atmosphere of Mars is mostly carbon dioxide and water vapour exists only in small quantities, their density is too small to produce the desired effect. But heating the surface, thus liberating some of the carbon dioxide locked away in the form of carbonates, may have some effect in increasing the atmospheric pressure and then the greenhouse effect may further increase the temperature. Many ideas have been put forward, from using large orbiting mirrors to dispersing a dark dust over the polar caps, from exploding thermonuclear bombs under the surface to sending an asteroid crashing onto the planet, or to digging very deep wells to extract heat from the inner part of the crust of the planet.

These methods are probably insufficient. However, by liberating chlorofluorocarbons (CFCs) into the atmosphere it could be possible to raise the temperature at its surface. Since the molecules of these gases are not entirely inert and are decomposed by light from the Sun, on a time scale of about 200 years, a continuous production of CFCs would be required to maintain a temperature higher than the present one. It has been estimated that an annual production of about 100,000 tonnes is needed to maintain the equatorial zone of Mars at a temperature similar to that in a temperate zone on Earth.[12] This is only a fraction of the amount produced annually on Earth in the 1970s, leading to the creation of the Antarctic ozone hole and to so much concern world-wide.

An initial production which is 10 times greater and 5,000 MW of electrical power, the output of five medium-sized power stations on Earth, is required to create the concentration of CFCs needed within 40 years. It appears then that it is feasible to heat the surface of Mars in this way, or, if a longer time, namely 80 years, is acceptable, this could be done by installing three medium-sized power stations and a plant equalling the CFC production on Earth in the 1970s. Clearly,

[12] R.M. Zubrin, D.A. Baker, Mars Direct, a proposal for the rapid exploration and colonisation of the red planet, in *Islands in the Sky*, Wiley, New York, 1996.

this is possible as the actual energy to heat the planet comes from the Sun and the role of humankind is that of tampering with the thermal regulation system of the planet. At a personal level, it is equivalent to changing slightly the setting of a domestic central heating system.

Perfluorocarbons (PFCs) could be used instead of CFCs. Their advantage is that they do not deplete ozone. So, as soon as some oxygen is formed in the Martian atmosphere, ozone may also start forming, and thus reduce the amount of ultraviolet light reaching the surface.

A number of changes would be triggered by a warming of the planet. The dry ice (solid carbon dioxide) in the polar caps would melt and carbon dioxide would be liberated, adding to that already present in the atmosphere. In about 40 years an atmospheric pressure equal to about one third of that on Earth could be reached. Then our colonial astronauts could each move around on the surface without a space suit, with just a mask and a bottle of oxygen, as scuba divers do.

The water in the permafrost would slowly melt, flowing over the surface and filling the beds of the ancient rivers and lakes, and forming polar caps of true ice, adding to the water ice already present at the poles. However, it is not known how long it would take for this to occur, as this depends mainly on how far below the surface the permafrost is. (At present there is no definite proof that a permafrost layer exists, even though there are strong hints in this direction.) The evaporation of water would start a fresh water cycle, with the formation of rain and snow. The increase of carbon dioxide and water vapour in the air, both aiding the greenhouse effect, would allow the production of CFCs to be reduced.

At this point, the planetary restoration would be complete and the planet would have regained its ancient aspect. As it took a very long time for Mars to lose its atmosphere and to cool down, humans could then limit themselves to producing the small quantity of CFCs required to stabilise conditions on Mars which, as past history has taught us, are naturally unstable.

It is still not known whether life developed when Mars was a warm planet rich in water, nor whether life is still present in some well-protected niche, having survived the catastrophic changes which upset the planet. But when Mars has been restored to its former glory (or even before then, as has already considered), humans will bring life from the Earth. In an atmosphere containing so much carbon dioxide vegetation would thrive, even more so if humans select or genetically modify certain species to make them more suitable for the Martian environment. It is likely that many plants and trees could be introduced early in the process, when the temperature and the pressure have increased only slightly above their present values. Agriculture would flourish in the plains and the rougher regions could be covered by forests. But plants have another important function – they produce oxygen. Originally, the atmosphere on Earth contained only carbon dioxide and nitrogen; oxygen was produced later by vegetation.[13] Exactly the same situation could occur on Mars, and it has been computed that it would not take more than a few centuries. Although this is a long time with respect to the human lifespan, it is nothing in terms of the geological history of a planet and not long even for the history of a civilisation. To obtain a Martian atmosphere which humans could breathe freely would be amazing.

This is, however, a controversial issue. To make the atmosphere breathable by humans the content of carbon dioxide must be reduced drastically and this would reduce the greenhouse effect too. Some hold that the choice will be between a warmer planet with an atmosphere in which humans must carry oxygen bottles and masks and a very cold atmosphere, with a low pressure but with more oxygen in it.

If these issues can be resolved satisfactorily, Mars could have a breathable atmosphere, a vegetation and fauna quite similar to the situation on Earth, even if adapted to the conditions of the planet Mars. It could support a population of several million, within a time that is similar to that separating us from the arrival of Christopher Columbus in America. Mars would then be the first planet terraformed by humans from the planet Earth. An interesting account of the efforts to

[13] Chapter 17 of J.I. Lunine, *Earth, Evolution of a Habitable World*, Cambridge University Press, Cambridge, 1999.

make Mars a planet similar to the Earth is discussed in works of science fiction.[14]

The scenario described above seems to be feasible. However, there are other avenues which could be followed to reduce the power needed to produce the CFCs. First, large mirrors could be put in orbit around the planet, which would heat the surface using reflected sunlight. This can be done at almost no cost if solar sails are used to reach the planet; once they have finished their task as propulsion devices they can be left in orbit and oriented in such a way as to melt the polar caps, liberating carbon dioxide into the atmosphere which, thanks to its greenhouse effect, would amplify the heating. If only a few spaceships were used, only a small effect would be obtained. Purpose-built large mirrors would be needed to obtain large-scale effects. They could be built in space, with aluminium mined on the Moon, or on some asteroids, and moved to Mars using the radiation pressure of sunlight.[15]

Alternatively, it may be possible to change the orbit of some small asteroids or comets and make them fall onto Mars. This could be done using relatively low thrust ion thrusters, and then the gravitational field of a giant planet to 'focus' them onto their target. The kinetic energy of the asteroid or comet would be converted totally into thermal energy, causing localised heating of the region around the point of impact and releasing the substances from which it is made. Comet nuclei, rich in ice, could increase the amount of water vapour in the atmosphere, while an asteroid rich in ammonia, if found, would liberate this gas into the atmosphere, increasing the greenhouse effect and creating a protective layer against ultraviolet light. It may seem a strange idea to project asteroids or comets into a planet to terraform it, but this actually mimics the natural mechanism which led to the formation of planets. It is likely that water and carbon, perhaps some

[14] K.S. Robinson, *Red Mars*, *Green Mars* and *Blue Mars*, Bantam Books, New York, 1993, 1994, 1996.

[15] R.M. Zubrin, C.P. McKay, Terraforming Mars, in *Islands in the Sky*, Wiley, New York, 1996.

already in the form of organic compounds, were delivered to the Earth by falling cometary nuclei.

The techniques described here to terraform Mars are perhaps too primitive to work precisely, but they are the result of just a few years of study on a completely new theme. They are based on the greenhouse effect and are an application to Mars of what has recently been learned about the effects of human activity on the Earth's environment. When it is time to pass from speculation to feasibility studies and then to actions, all these subjects will be known in much greater detail. And it is possible that simpler, more effective, and perhaps more economical, methods will be found. Nanotechnologies,[16] for instance, promise revolutionary applications in this field as in many others. Molecular automata, with their incredible ability to replicate themselves and then to operate on a very large scale with negligible costs, could change the chemical composition and the characteristics of a planetary atmosphere in times which are a fraction of those mentioned above. Here, the aim is more to show that it should be possible to terraform Mars rather than to describe the detailed ways in which technical problems can be solved.

To terraform Mars will be a huge undertaking, greater than some engineering feats of the past, like building the Great Wall of China or the Suez Canal. Apart from giving humanity a new planet on which to settle, it is bound to yield an invaluable body of scientific and technological knowledge. One important fall-out from this will be a detailed knowledge of the mechanisms which regulate a planetary environment. Such knowledge is definitely needed as humankind has, in the last century, acquired the ability to change its own ecosystem and now must understand the consequences of its actions. Moreover, this exercise will be very useful in the future. If the final frontier of the title is to be extended further, other planets will need to be terraformed, but under even more difficult circumstances.

[16] K.E. Drexler, *Engines of Creation*, Anchor Press, New York, 1986.

A problem which is not technical remains. Even assuming that terraforming a planet is feasible, is it advisable or morally acceptable? But first one objection must be removed, concerning the massive use of greenhouse gases, particularly CFCs. While the production of CFCs is now banned on this planet, their effects elsewhere in the solar system could be beneficial. What could irretrievably ruin the Earth's atmosphere, making a hell – like Venus – out of it, may be able to return to Mars its ancient grandeur. Is the terraforming process worthwhile? And acceptable? In his book *The Search for Life on Other Planets*,[17] Bruce Jakowsky lists the following points in relation to terraforming Mars and introducing an active biosphere there. Seven arguments in favour are:

1. A thick, nonbreathable atmosphere of carbon dioxide would simplify life for future astronauts or colonists; they could explore the red planet equipped only with breathing apparatus rather than with full environmental space suits.

2. Locally generated biomass would be an important source of energy, food, and other useful materials for astronauts or colonists.

3. Such an activity would provide a long-term challenge on which humans could focus, with a goal that is both useful and desirable for humans.

4. Such a project would be an essential prerequisite to any future human colonisation of Mars.

5. An active biosphere on Mars would provide a refuge for many forms of life on another planet in the solar system in the event of war or natural global catastrophe that might destroy life on Earth.

6. Much of the research would be highly relevant to addressing environmental problems on the Earth and to understanding its biosphere.

[17] B. Jakowsky, *The Search for Life on Other Planets*, Cambridge University Press, Cambridge, 1998.

7.   Solar system developments are far less threatening than military developments or an arms race on Earth, and would provide an outlet for international co-operation/competition and/or technology developments.

And seven arguments against are:

1.   The time scale is too long compared with the lifetimes of governmental institutions to maintain the necessary commitment to such a project.
2.   It is not clear that there are significant economic benefits, especially in the short term, that would be commensurate with the cost and effort involved.
3.   Scarce human and economic resources would be diverted from other worthwhile projects, such as addressing social and terrestrial environmental problems.
4.   Something could go wrong in the course of the project that could damage the new Martian biosphere beyond repair, leaving us worse off than before.
5.   Humans have done such a bad job of managing the Earth's environment that it is presumptuous to imagine that they can be wise enough to succeed on another world.
6.   If terraforming were successful, Mars might become a tempting target for military and/or economic exploitation; this could generate more sociopolitical problems than we have at present.
7.   The evolution of a Martian biosphere could be inherently unpredictable, and might be detrimental to humans or to Earth.

An interesting general observation about the idea that humankind might introduce changes to the Martian environment was made by McKay and Haynes:[18] 'If and only if no potentially viable forms of life are found should we attempt to introduce immigrant species from Earth. ... What would be the greater good, Mars barren

[18] C.P. McKay and R.H. Haynes, Should we implant life on Mars? *Scientific American*, page 108, December 1990.

or Mars endowed with life? . . . Should the Martian biosphere be tended to ensure at least early development in a manner agreeable to *Homo sapiens*?'

The Earth's (or the Martian) ecosystem is the result of a large number of factors playing different and contrasting roles. One of these factors is humankind: the fact that human beings are intelligent and conscious does not deprive them of their right to play a part in the game. However, that places upon humankind the burden of behaving wisely, trying to predict the consequences of its action. But this responsibility must not paralyse us – to paraphrase the words of Hamlet:[19] 'this enterprise of great pitch and moment . . . to lose the name of action'.

The first planet of the solar system to have been terraformed was the Earth, long ago, by the living beings who changed an atmosphere based on carbon dioxide to one rich in oxygen. To wonder whether they had a right to do that is clearly meaningless.

If there is no life at present on Mars, all the ethical problems may seem to stop with the question of whether humankind has the right to play some active role in shaping that infinitesimal part of the Universe over which it may exert some influence. But if living beings are discovered on Mars the whole issue must be studied from two points of view. How might the terraforming operations affect them, and what would be the consequences of their presence for the whole project? The changes might make them evolve, and perhaps they would thus complete a process which was stopped in the distant past. But the changes imposed by humans are likely to be too sudden to allow living beings to become adapted to the new conditions and the risk that they will not survive is great. Thus, it is important that the lifeforms present on Mars, if any, are studied in great detail before starting any work aimed at changing the planet. All the practical measures necessary to protect and preserve what can only be considered as an extreme and special case of biodiversity must be taken.

---

[19] W. Shakespeare, *Hamlet*, Act III, Scene 1.

A radical position is based on the observation that, even if some living beings still exist on Mars, there is no doubt that they failed to produce an extensive biosphere on that planet. To substitute them with more successful lifeforms which could spread over the whole planet is an action perfectly in line with the basic logic of evolution – the fittest forms of life survive and propagate.

These issues are useful to start a debate. And since terraforming Mars is a distant prospect, there is plenty of time to acquire a better knowledge and a deeper understanding of all the issues involved.

7   Exploitation of the solar system

The robotic exploration of the solar system has been outlined in Chapter 4, while the possibilities for human beings exploring Mars and then colonising it have been dealt with in Chapter 6. Owing to the complexities, costs and potential dangers of sending human crews to more distant destinations in the solar system, this will only happen when robots show that such expeditions are justified. Are Mercury and Venus likely to be useful for the human species? What valuable minerals are present on asteroids which we could exploit? What could we extract from the giant gas planets or their rocky satellites? And are there any useful resources in the Oort cloud?

THE INNER PLANETS: MERCURY AND VENUS

The planet Mercury[1] is an extremely hostile environment for humans. Its rate of rotation is very slow, with a period of about 60 Earth days. It has no atmosphere and its general aspect reminds us of the Moon (Figure 7.1 and Figure 7.2). Its surface has temperatures ranging from 350 °C on the side facing the Sun to −170 °C during the night. However, space probes have found water ice in its polar regions. Notwithstanding the extreme environmental conditions, there are some scenarios in which human settlements could be set up on this planet.

One may first ask, and not without reason, why humans might be interested in visiting or settling in such a place. The resources of Mercury are two-fold – solar energy, available in huge

---

[1] See F.W. Taylor, *The Cambridge Photographic Guide to the Planets*, Cambridge University Press, Cambridge, 2001.

Figure 7.1. The planet Mercury as observed by *Mariner* 10 in 1974 (NASA photo).

quantities, and metals, particularly aluminium. Being the planet closest to the Sun, Mercury would be quite easy to reach using solar sails.

Mercury could be a good place to build solar power stations. The energy so produced could be used on the spot, to extract and to transform metals, or for launching interstellar probes with laser sail propulsion (see Chapter 9). As far as its metal resources are concerned, the density of Mercury is very high for a planet of its size. This suggests that it is not made just of silicates, like the Moon, but has a core of heavy elements as the Earth does. Metals would thus be expected to

Figure 7.2. A close-up view of the surface of the planet Mercury (NASA photo).

be quite common on Mercury, although it is not yet known whether they are only in the inner core or also on the surface. And it is still not clear whether it would be more convenient to get them from some asteroids.

It is not technologically difficult to build a permanent base on Mercury, but it is difficult to find a good reason to do so at present. Perhaps the only reason would be to create a base where those involved in the construction and maintenance of a large solar power station, relatively close to the Sun, could live and be shielded from the Sun's activity. An artist's impression of the planet Mercury is shown in Figure 7.3.

Neither does Venus seem to be an interesting place to settle. Its size is close to that of the Earth, but the similarity between the two

Figure 7.3. An artist's impression of the desolate surface of Mercury (NASA image).

Figure 7.4. Venus, as seen by the *Magellan* probe using its synthetic aperture radar (NASA photo).

planets ends there. The planet is always covered by thick layers of clouds and its surface cannot be seen from space by optical observation. However, the *Magellan* probe has mapped its surface accurately[2] using a synthetic aperture radar (Figure 1.22 and Figure 7.4). The results have been turned into a three-dimensional model over which a virtual aircraft can be flown. The surface of Venus is a true hot hell, with temperatures reaching 500 °C. The atmospheric pressure is also very high, about 90 times that on the Earth's surface. Its

[2] See R. Greeley and R. Batson, *The Compact NASA Atlas of the Solar System*, Cambridge University Press, Cambridge, 2001.

rate of rotation about its axis is extremely slow, the Venus day being about 243 Earth days long.

While it will be possible to land humans and even to build a permanent base on Mars before any operation to alter its natural conditions are attempted, on Venus things are much more complicated. From a strictly technical point of view, it may be possible to land a crew there, but the extreme environment would make life impossible. On Venus, terraforming operations would have to be carried out before, and not after, any attempt at manned exploration.

The present environment on Venus is believed to be the outcome of a runaway greenhouse effect, the dangers of which we, who live on the Earth, should heed. An increase of the Venusian temperature caused the evaporation of the seas, if they ever existed, and the production of carbon dioxide from the carbonates in the soil. The increasing amounts of water vapour and carbon dioxide in the atmosphere caused a further increase of the greenhouse effect, and the very high temperatures which we now observe. The water vapour was decomposed by the Sun's light into oxygen and hydrogen, the latter light gas disappearing into space. Venus does not have more carbon dioxide than the Earth, but it is all in the atmosphere instead of being fixed in the soil and absorbed in the oceans.

To terraform Venus carbon dioxide in the atmosphere would have to be removed, chemically transformed into solid carbonates and put into the ground. Doing this would reduce the temperature, but it is impossible to reduce the temperature unless the carbon dioxide disappears, and it is impossible to remove the carbon dioxide unless the temperature is reduced.

Adding vegetal lifeforms would not help; algae or other microscopic organisms liberated in the atmosphere could decompose carbon dioxide into oxygen and carbon, but these would recombine, owing to the high prevailing temperature. Better results could be obtained by using light metals, such as calcium or magnesium, which react with carbon dioxide to produce carbonates. But the quantities of metal required are billions of billions of tonnes, of the same order as the mass of one of the biggest asteroids in the solar system.

We might contemplate mining huge quantities of calcium and magnesium from Mercury, and then sending them to Venus using electromagnetic launchers. Theoretically this could succeed, as the operation only requires very large quantities of energy, which are available on Mercury. However, the kinetic energy of the pellets launched to Venus would heat the planet, rather than cool it as is desired.

A possible way of breaking the high temperature – high carbon dioxide content cycle is to deploy a very large reflecting surface (a sort of huge solar sail) between Venus and the Sun to cool the planet. Slowly the temperature of the atmosphere would decrease, causing a decrease of the amount of carbon dioxide; in turn this would reduce the greenhouse effect, leading to a further decrease of the temperature. Seeding the upper atmosphere with algae could hasten the process. The size of an 'astro-engineering' project of this type is huge, but no more so than the construction of lenses of about 1,000 km in diameter needed for interstellar probes propelled by laser sails (see Figure 8.5).

MINING BASES IN SPACE: THE ASTEROIDS

Away from the Earth, beyond the orbit of Mars, there is the asteroid belt.[3] Asteroids are rocky bodies with a wide range of sizes, made of material which did not participate in those processes which formed the planets of our solar system 4.5 billion years ago. The chemical composition of the asteroids is fairly well known, and they have been subdivided into various classes according to the elements and chemical substances which they contain. The study of meteorites which have arrived on the Earth's surface has shown that they contain the same substances and today it is certain that most of them have come from the asteroid belt.

---

[3] The average radii of the orbits of Mars and Jupiter around the Sun are about 1.5 and 5.2 AU. The asteroid belt stretches between 2 and 4 AU: its width is thus the diameter of the Earth's orbit around the Sun.

Many of the minerals which can be found on the asteroids have a large economic value. To make profits as well as to replenish the Earth's dwindling resources of some precious metals, mining the asteroids might take place in the more-or-less distant future. How far in the future will be determined by the cost of extraction of such minerals from the asteroid belt being competitive with their cost of extraction from other places, the Earth, the Moon and Mars.

Most asteroids have a composition similar to that of the Earth's crust – mostly silicon, iron and calcium oxides. About 20% of the asteroids are carbonaceous chondrites, made of carbon, hydrogen, nitrogen and other volatiles, among which there is water. Even in 1905, Robert Goddard suggested using water from the asteroids to obtain hydrogen and oxygen as rocket propellant.

The asteroids which might become the most important sources of raw materials are the iron asteroids. They are relatively rare, constituting 3% of the total number of asteroids, but their total number is so high as to contain a huge quantity of iron and other metals, like nickel and chromium and also the precious metals, platinum and gold. Their economic, and also their technological, value is huge. A small iron asteroid, 1 km in diameter, would contain about 10 billion tonnes of iron, the present Earth's production in about 12 years, 1 billion tonnes of nickel, 100,000 tonnes of platinum and 10,000 tonnes of gold. At the present values of these metals, this amounts to something like US $1,000 billion.[4] The quantities of these metals present in the asteroids are so large that it is possible that their exploitation could lead to a decrease in their cost, so that precious metals could have more technological applications (e.g., gold as an excellent electrical conductor).

Near Earth asteroids (NEA) are easier to reach than the main belt asteroids, and it is likely that they will be the first ones to be exploited. It is even easier to reach them than it is to reach the Moon; to land on their surface in a very weak gravitational field is straightforward, as the successful landing of the *NEAR* probe has shown. To exploit

---

[4] N. Prantzos, *Our Cosmic Future*, Cambridge University Press, Cambridge, 2000.

NEA has an added benefit. It is the best way to defuse the danger of a collision of such an asteroid with the Earth.

Exploitation of the asteroids is a very common theme of science fiction and many scenarios have been devised. Here, two possibilities are mentioned – bases on the asteroids themselves and space habitats located near them. In the first case these will be similar to lunar bases (see Chapter 5), apart from the fact that the gravitational field of any asteroid is much smaller than that of the Moon. In the second case the space habitats would not be different from those described for the settlement of circumterrestrial space (see Chapter 3).

The main advantage of the asteroids is their very low gravitational field, which makes it relatively easy to send the materials mined on them to almost any destination in the solar system. The transportation costs to the Earth's orbit or to any space habitat depend mostly on the energy needed to reach the heliocentric orbit of the destination from that of the asteroid. Very low gravity has its disadvantages too, first for the people who would live on an asteroid base. Secondly, owing to the low gravity, asteroids are more like piles of rubble than solid bodies and they are supposed to be very brittle. It would thus be very difficult to work on them, for example to dig the cave dwellings that often feature in science fiction. The results obtained by the *NEAR–Shoemaker* probe on the asteroid Eros, however, seem to show that Eros is not a 'rubble pile' but a solid celestial body. Perhaps it is a better idea not to live on the asteroid, but in a space habitat located nearby (Figure 3.10).

Another possibility could become important in the more distant future. The robotic exploitation of the asteroid belt, using large robots which only occasionally had to be attended by maintenance crews, is foreseen. In this scenario the number of humans living in space beyond the orbit of Mars would be very small.

ENERGY FROM THE GIANT GAS PLANETS

Potential resources located beyond the orbit of Mars are not limited to the minerals of the asteroid belt. The giant planets, particularly Jupiter and Saturn, are rich in the different isotopes of hydrogen and helium; these would become strategic resources as soon as thermonuclear fusion is tamed. Deuterium, an isotope of hydrogen, and helium 3, very rare on the Earth, may come to have the role that oil has nowadays. It has been calculated[5] that, with the helium 3 which is in Jupiter's atmosphere, it would be possible to obtain 5,600,000,000 TW years.[6] This is an enormous figure compared with the existing energy resources on Earth, evaluated as 3,000 TW years, or with the energy used in just one year, 1992, equal to 12 TW years. But if humankind goes on increasing its energy consumption exponentially, the exploitation of energy sources of extraterrestrial origin, and not only of raw materials, will be essential in 200 years' time.

The extraction of the resources of the giant planets can be achieved only using robotic systems, as humans cannot live in such large gravitational fields as exist at the apparent surface of these gas planets. However, their satellites could be settled. All the giant planets of the solar system have many satellites and a more-or-less complex system of rings consisting of dust particles. Jupiter has 16 known satellites and a single, very thin, ring; Saturn has 18 satellites and seven rings, two of which are very large and spectacular. Uranus has 15 satellites and 11 thin rings and Neptune has eight satellites and six rings. Many of the satellites are small and irregular and look like asteroids captured by the strong gravitational field of such a huge planet. Some satellites, termed 'shepherd' satellites, keep the rings in place with their gravitational pull (Figure 7.5).

Other planetary satellites are large and essentially spherical, of the size of the Earth's Moon or larger. They have the characteristics of small planets, Ganymede and Titan being larger than Mercury (see Table 7.1: the well-known Uranian satellite Miranda is not included owing to its small size). The planetary satellites are very different

[5] R.M. Zubrin, Colonising the outer solar system, in *Islands in the Sky*, Wiley, New York, 1996.
[6] 1 TW (terawatt) of power is equal to one million kW; an energy consumption of 1 TW year is then equal to 4.4 billion kW hours. The power of sunlight incident on the Earth's atmosphere is, for comparison, 10,000 TW.

Figure 7.5. Diagram of the rings of Saturn with the 'shepherd' satellites (NASA diagram).

from each other, some being frozen worlds, with a solid surface mostly made of ice, not only water ice but other solidified gases, while others are very hot, due to the strong tides produced by the combined gravitational pull of the planet and other satellites. Europa (Figure 4.14) is thought to have an ocean of liquid water (one essential for most forms of life) under its ice surface. By contrast, Titan has a very thick and cloudy atmosphere, mostly nitrogen. It may have seas of liquid methane, which might be exploited as an energy resource or a raw material for chemical industries.

It is important to think of exploiting the resources of the giant planets and their satellites. There could be human outposts or even colonies on some of the satellites, perhaps Europa, Io, Titan or Miranda. The image of Saturn, with three of its satellites taken by *Voyager 2* while near Titan (Figure 7.6), gives an idea of the fantastic sight which the first settlers of a base on that satellite could admire.

Table 7.1. *Diameter and average distance from the centre of the planet of the main satellites of the giant planets, compared with the Earth's Moon. Only those satellites having a diameter of at least 1000 km are listed*

Planet	Satellite	Diameter (km)	Distance (km)
Jupiter	Io	3,630	421,600
	Europa	3,138	670,900
	Ganymede	5,262	1,070,000
	Callisto	4,800	1,883,000
Saturn	Tethys	1,048	294,660
	Dione	1,118	377,400
	Titan	5,550	1,221,850
	Iapetus	1,436	3,561,300
Uranus	Ariel	2,500	190,900
	Umbriel	1,158	266,000
	Titania	1,169	436,300
	Oberon	1,578	583,400
Neptune	Triton	2,705	354,800
(Earth)	(Moon)	3,400	384,000

THE FRONTIER OF THE SOLAR SYSTEM

Some scientists, such as the physicist Freeman Dyson, have gone as far as imagining human settlements even beyond the giant planets, in the outer reaches of the solar system.[7] The asteroids and the cometoids of the Kuiper belt and maybe even the comet nuclei of the Oort cloud could be settled. There is no doubt that, when our technology succeeds in carrying human beings to such an enormous distance, it would also be possible to build permanent settlements in a zone so far from the Sun. This frontier of the solar system is really vast (see

[7] F. Dyson, *Infinite in All Directions*, Harper and Row, New York, 1988.

Figure 7.6. Saturn and its rings as seen by the *Voyager 2* probe when it was close to its satellite Titan. At the bottom are the satellites Tethys (whose shadow can be seen on the planet), Dione and Rhea (NASA photo).

Chapter 8). The Oort cloud is densest about half a light year from the Sun (about 32,000 Astronomical Units), and contains perhaps 2,000 billion cometary nuclei. The outer reaches of the Oort cloud are at a distance of 100,000 Astronomical Units, about one and a half light years, from the Sun. If our nearest neighbouring stars were surrounded by a similar cloud of cometary nuclei, their outer regions would almost touch each other.

The Oort cloud,[8] very rich in water, organic matter and other raw materials, is a valuable resource. A civilisation could, perhaps,

[8] The Oort cloud, and many fascinating objects in the solar system, are interestingly discussed by S.R. Taylor, *Destiny or Chance, Our Solar System and its Place in the Cosmos*, Cambridge University Press, Cambridge, 1998, and by H.Y. Sween Jr., *Meteorites and their Parent Planet*, (second edition), Cambridge University Press, Cambridge, 1998.

use nuclear fusion to settle this extremely dim frontier of the solar system. Small nomadic communities, each living in a self-sufficient space habitat and moving from one cometary nucleus to another, have been described. They are the solar system equivalent of those on planet Earth who have settled Lapland, Greenland or Alaska. Human colonies living in that extreme region will find it very difficult to – or may even never want to – go back nearer to the Sun, to be in a much stronger gravitational field and to breath a dense atmosphere, just as an Inuit may be unlikely to enjoy – or even wish for – a visit to New York.

8 Beyond the Pillars of Hercules

Space is infinite, almost. The nearest stars are several light years from us, and our galaxy, the Milky Way, is a hundred thousand light years across. To travel to possible planetary systems around other stars, let alone beyond the galaxy, we require radical new propulsion technologies which may only become available in the far future.

Should human beings aim to visit other worlds in space? Or would it be better to send intelligent, self-reproducing probes instead? Or should we be content to explore space only in the mode of virtual reality?

HUGE DISTANCES, YET INSUFFICIENT SPEED

The ancient Greeks located the limits of their navigable waters, the Mediterranean Sea, at the Pillars of Hercules, i.e. the mountains on either side of the Straits of Gibraltar. Even if a few ancient sailors actually ventured beyond the Mediterranean, the Pillars of Hercules remained the symbol of an impassable limit.

Although a few robotic probes might, in the future, leave the solar system bound for nearby extrasolar planets, are there Pillars of Hercules for space travellers in the outer reaches of the solar system? Do they preclude us for ever from reaching for the stars?

The distances which separate the stars of our Milky Way galaxy are truly vast, at least when we compare them with distances which are familiar to us in everyday life. How we perceive distances depends on our habits and on the technologies available to us. A 'cave man' would, no doubt, have defined as far away a place 10 km from his dwelling, while today's business executive, used to travelling by plane, thinks all major European cities are quite near to each other. But interstellar distances are in a completely different realm.

The star nearest to us is Proxima Centauri, a small red dwarf belonging to the triple star Alpha Centauri. It is 4.3 light years (about 273,000 Astronomical Units (AU[1]), or 41 million million km) away, while the other two components are slightly further, at 4.4 light years. If we build a model with the Sun being represented by a ball with a diameter of 1 cm at the centre of a football field, the Earth would be as small as the point of a pin with a diameter of 0.1 mm at a distance of 1 m from it. The orbit of Pluto would fit almost exactly between the two goals, and Proxima Centauri would be almost 300 km away. The *Voyager 1* probe, which left the solar system at a speed of 3.5 AU/year, i.e. 16 km/s, will reach a distance of 4.3 light years in about 80,000 years. It will not pass near Proxima Centauri, as it is speeding away from us in another direction. *Voyager 2* and the two *Pioneer* probes (Figure 8.1) are travelling more slowly; none of them will pass close to a star in the next million years.

Within a range of 10 light years there are seven stars, two of which are double stars, including Sirius, and one a triple star system. The number of stars grows with increasing distance. Within 20 light years there are 68 and many hundreds in a range of 30 light years. Figures of this type soon become uncertain, as the distance from the Sun to different stars of our galaxy is not accurately known. However, thanks to a space mission, the astrometric satellite *Hipparcos* (Figure 8.2), stellar distances may now be measured with good precision. We shall discuss later how many of these stellar systems might also have planets.

To launch a probe or a spaceship towards a nearby star in order to satisfy our human curiosity by visiting its planetary system, a speed far greater than those typical of current methods used for exploring

---

[1] One Astronomical Unit is the average distance from the Sun to the Earth, about 150 million km.

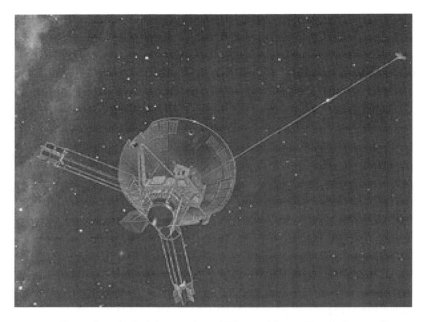

Figure 8.1. Artist's impression of *Pioneer 10* travelling in interstellar space (NASA drawing).

our solar system must be reached. But there is a physical limit to the maximum speed which any projectile can attain – the speed of light. Relativity theory states that no material object can travel at, or faster than, the speed of light;[2] this limit also holds for any means of transmitting information. The speed of light is thus a cosmic speed limit, and all our dreams of interstellar exploration must take this fact into account.

The theoretical minimum travel time to any destination is, in years, equal to its distance in light years. If a probe could be launched, almost at the speed of light, towards Alpha Centauri with incredibly advanced technologies, we could receive the first images of that star only after 9 years – 4.5 years for the journey and 4.5 years for the

---

[2] The speed of light is generally indicated with the symbol c; its value in a vacuum is 299,792.458 km/s, or, to a good approximation, 300,000 km/s.

Figure 8.2. Artist's impression of the *Hipparcos* astrometric satellite (ESA drawing).

transmission broadcast by the probe to reach us. Even if a rocket could launch a space probe at 3,000 km/s, one hundredth of the speed of light, the journey to our nearest star would take 430 years.

The speed of the *Pioneer* and the *Voyager* probes leaving the solar system is of the order of half of one ten thousandth of the speed of light ($v = 0.00005c$). Advanced current technology might allow us to leave the solar system with a speed (in the heliocentric frame, i.e. with respect to the Sun) of about one ten thousandth of the speed of light ($v/c = 0.0001$), i.e. 30 km/s. Then we could reach the nearest star in about 43,000 years, an impossibly long journey time for the human race to contemplate.

Future interstellar travel is conventionally be subdivided into four categories, depending on the average speed of the spacecraft:

- *Slow interstellar travel.* At a speed of 1% of the speed of light ($v = 0.01c$), the time needed to reach the nearest stars is about 500 years, but the energy required is 10,000 times greater than that available using current advanced chemical propulsion systems.
- *Fast interstellar travel.* The speed is of the order of 10% of the speed of light ($v = 0.1c$). The time required to reach the nearest stars is about 50 years, but the energy supplied must be increased a further hundred fold.
- *Relativistic interstellar travel.* The speed is close to the speed of light (v is between $0.8c$ and c). The time needed to reach the nearest stars is about 5 years, but relativistic effects start to be felt; one of their consequences is that aboard the spacecraft time slows down, and the journey time for the humans and machinery aboard is shorter than this. The closer the speed is to the speed of light, the shorter the journey will be, in terms of on-board time.
- *Superluminal (FTL: Faster Than Light) interstellar travel.* The speed is greater than the speed of light ($v > c$). According to relativity theory, this type of interstellar mission is impossible (but see Chapter 10).

But speed in itself is not everything. Huge amounts of energy are needed to accelerate any object to the speed of 3,000 km/s, classified here as slow interstellar travel. And from the theory of relativity we know that an infinite amount of energy is needed to accelerate a body to the speed of light; this accounts for the impossibility of travelling as fast as light. Even to reach a speed of one hundredth of the speed of light, when relativistic effects can still be ignored, the amount of energy that has to be supplied is huge. If the mass of the spacecraft were one tonne, the energy required to accelerate the payload to $v/c = 0.01$

would be 4.5×10^{15} joules,[3] an enormous amount of energy, which is comparable to the amount of energy consumed by a European country in one day, and at a prohibitive cost. Moreover, in a mission of that kind the probe would remain in the Alpha Centauri system (provided that the star has a planetary system) for only a few days. If the spacecraft had to stay longer near its target, a braking manoeuvre, requiring another 4.5×10^{15} joules of energy, would have to be performed. Such simple, yet fundamental, calculations lead many to doubt that humans could ever cross the interstellar gulf, either in person or using robots. And yet the International Academy of Astronautics has a committee dealing with interstellar space exploration.

The last three chapters of this book discuss the perspectives that interstellar flight could open up to humankind, and possible ways of pursuing them. They consider the 'encounters' humans could make, and must be ready to make, if they ever decide to begin this ambitious enterprise, and make a commitment to space as the final frontier.

THEORETICAL AND PRACTICAL IMPOSSIBILITIES

Hermann Oberth concluded his book *Ways of Space Navigation*, written in 1929, with the sentence: 'Nothing is impossible in this world; it is enough to find out the means through which it can be made'. Or, to repeat another well-known quotation of Konstantin Tsiolkovsky: 'The impossible of yesterday is the hope of today and the fact of tomorrow'. Such mantras come from an over optimistic attitude. They do not always hold, and there are some things that are impossible today, tomorrow and for ever.

At this point we make a distinction between theoretical impossibilities and practical impossibilities. To the first group belong those

---

[3] This calculation refers to the kinetic energy which the probe must acquire to reach this velocity, and which must be imparted by the propulsion system. This type of calculation is exactly that used by A.W. Bickerton in 1926 to show that it was impossible to 'shoot at the Moon' (see Chapter 1). He worked out the kinetic energy needed to reach the escape velocity, and thought that it was impossible to impart this amount of energy to any object. We can hope that future technological advances will again circumvent this reasoning.

impossibilities which follow from a sufficiently validated scientific theory. A very good example of a theoretical impossibility is perpetual motion, which goes against the laws of thermodynamics. To the second group belong those things which humans never succeeded in doing in the past and cannot do today owing to insufficient technological or economic means. A good example is the flight of heavier-than-air machines until the last part of the nineteenth century. The materials, engines and theoretical knowledge needed to build an aircraft were not there, but the flight of birds was a practical demonstration that it could be attempted.

This classification is not really clear cut for all time, and some impossibilities can move from one class to the other. One scientific theory can be replaced by another – more general – one, and something which the first demonstrated to be impossible can pass through 'holes' opened by the second. Alternatively, something which is just difficult may seem to be theoretically impossible in the light of a poorly interpreted, or wrong, theory.

Impossibilities have been common in the history of technology. For example, technology historians divide the energy sources used by humankind into five types. Striking changes occurred in material culture, and also in culture in general, each time we shifted from one to the next.[4] In the first stage we had only our own muscular strength at our disposal. The second stage, which started with the taming of a few animal species during the neolithic revolution, is characterised by the availability of power generated by domestic animals (e.g., the use of oxen to pull ploughs). The third stage, which began with the introduction of the water wheel in the last years of the Roman Empire, is based on running water and wind as energy sources. The fourth stage began with the industrial revolution in the eighteenth century and is based on burning fossil fuels. The fifth is characterised by nuclear energy: it is likely that it will start on a larger scale only when controlled nuclear fusion becomes possible.

---

[4] R.J. Forbes, Prime movers, in *A History of Technology*, Vol. 2, C. Singer *et al.* (editors), Clarendon Press, Oxford, 1958.

Today, we take for granted the availability of energy at the flick of a switch. Just a few generations ago that would have been considered impossible. An example can demonstrate the extreme dearth of energy, by present standards, in which humankind lived until two centuries ago. The mechanical power generated by a water wheel of about 5 m diameter is only a few kilowatts, and that of a large windmill up to 10 kW. The *Domesday Book*, the survey of Britain made for William the Conqueror in 1086, reports that in England, south of the rivers Trent and Severn, there were 5,624 water wheels. Thus, in the most populated part of England the total power available was perhaps no more than 20 MW. Such power is only a fraction of the power at the fingertips of the pilot of a Boeing 747 (150 MW) travelling at 900 km/ hour with a thrust of 600 kN. A modern power station generates up to 2,000 MW.

The present energetic impossibility of making an interstellar journey is with reference to a civilisation whose energy sources are those of the fourth stage, namely fossil fuels. Nuclear fusion – the source of solar energy – or even more advanced forms of energy are the requisite enabling technologies for travelling to the stars.

INTERSTELLAR PROPULSION

The chemical, nuclear or electric rockets mentioned in previous chapters are insufficiently powerful to propel interstellar probes. Their exhaust velocity, typically 10 km/s, and with a maximum value of 100 km/s, is totally unsuitable for interstellar journeys with a duration less than many millennia. Radically different and innovative launch technologies[5] must then be designed, developed and tested, first on the ground and then in space. Several of these are evaluated briefly here.

Nuclear propulsion

A mix of deuterium and helium 3 (about 1 cm^3), ignited hundreds of times per second by laser beams, could produce exhaust velocities in

---

[5] E. Mallove and G. Matloff, *The Starflight Handbook*, John Wiley & Sons, New York, 1989.

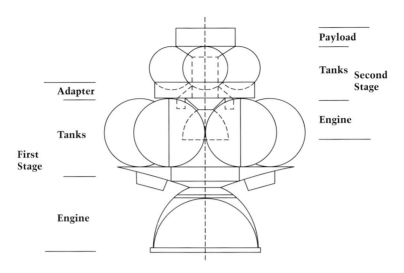

Payload

Tanks   Second
        Stage

Adapter

Engine

Tanks

First
Stage

Engine

Figure 8.3. Sketch of the two-stage *Daedalus* nuclear spacecraft.

the range of 10,000 km/s, about 3% of the speed of light. One study on pulsed nuclear propulsion was the Daedalus project, developed between 1973 and 1978 by some members of the British Interplanetary Society. It was a feasibility study for a giant two-stage probe, with a mass of 450 tonnes, propelled by a pulsed nuclear system with a specific impulse (see Appendix C) of about one million seconds. Starting with an acceleration of about 1.5% of 1g, produced by 250 small nuclear explosions per second, the first stage would have to work for 750 days before being discarded. Then the second stage would be fired up for another 640 days until, nearly 4 years after launch, the cruise speed of 12% of the speed of light was reached. The total mass at launch had to be 54,000 tonnes, 50,000 tonnes of which was made up by the deuterium and helium 3 propellant (Figure 8.3).

The study was for a one-way journey to Barnard's star (5.9 light years away, the nearest star which is not a multiple star system) and its hypothetical planetary system,[6] which could be reached in

---

[6] At the time of the development of the Daedalus project Barnard's star was believed to have a planetary system. The discovery of a planet orbiting Barnard's star was

50 years. It was an ambitious mission, in which the very large quantities of helium 3 needed had to be 'mined' from the atmosphere of Jupiter.

Antimatter propulsion

If a small quantity of antimatter is put into contact with an equal amount of normal matter, they immediately combine and are transformed completely into energy. The idea of using matter/antimatter annihilation for space propulsion was put forward by Eugene Sänger in the 1950s. His idea was to use electrons and positrons which, annihilating, produce a beam of gamma rays, i.e. high-energy photons. The photons would generate a thrust – this is the concept of the photon rocket. The highest possible ejection velocity, namely the speed of light, is obtained.

It now seems that this is not feasible. The development of antimatter rockets may be linked with the use of antimatter, mostly antihydrogen (antiprotons), reacting with normal hydrogen. The result of this annihilation process is the production of π-mesons, which produce gamma rays and neutrinos. The thrust could be obtained directly by using the beam of π-mesons and photons, or they could heat some inert propellant which then produces a high-speed jet. There would be enormous difficulties making a propulsion device of this type, one problem being how to contain antimatter and another the cost. At present tiny quantities of antimatter are produced in the

published in 1916. The astronomer Peter van de Kamp, of the Sproul Observatory, worked on this topic all his life, taking as many as 2,000 photographic plates of the star in the years from 1938 to 1962; in 1975 he concluded that the star has at least two planets, with masses of 0.4 and 1.0 times the mass of Jupiter. Other astronomers performed measurements on plates taken using other telescopes; finding no trace of these planets they put forward the hypothesis that the apparent motions of Barnard's Star were due not to the presence of planets, but to optical errors of the telescope originally used. Van de Kamp continued to believe that the planets he discovered actually existed until his death, in 1995, but few other astronomers agreed with him. After more than 80 years of research, the existence of planets around Barnard's Star, one of the stars closest to us, is more uncertain than ever.

world's largest particle accelerators. While being sufficient for devices like ICAN (Figure 4.8), they are far too small for primary propulsion of an interstellar probe. The present cost of antimatter is about US $100 billion for a milligram; such costs must be reduced by at least one order of magnitude before antimatter propulsion can be considered a viable competitor with chemical propulsion.

Solar sails

Solar sails (described in Chapter 4) could be the propulsion devices suitable for future interplanetary navigation. They could accelerate a probe to a fairly high speed within the solar system, and show an improved performance if the mass per unit area of sail surface could be decreased. A solar sail, made of an ultra-thin aluminium foil backed by an ultra-thin plastic layer, could take a probe out of the solar system at a speed greater – but not much greater – than that of *Voyager*.

The lighter the sail, the higher is its acceleration. Advanced sails (without a plastic backing) made of ultra-thin aluminium film, and backed by an even thinner chromium layer which helps in radiating heat to space and keeping the sail cool, have been proposed. The difficulties of handling and deploying such a very lightweight structure are great, but their performance is much improved.

Laser sails

If, instead of using sunlight, a well-collimated laser beam were to exert its radiation pressure on the sail for the whole of an interstellar journey, great improvements over a solar sail would accrue. Very large pieces of equipment would be needed – a very powerful laser, possibly powered by solar energy, and a huge lens to focus the beam. For a large probe a set of solar lasers with a power of millions of megawatts at the orbit of Mercury and a lens with a diameter of some thousands of kilometres, made of a very thin transparent plastic film, near the orbit of Jupiter would be required (Figure 8.4). Once the infrastructure had been built, no mean feat, it could be used to launch many

Figure 8.4. Artist's impression of a laser sail propelled by a laser powered by solar panels (NASA image).

probes, each travelling at speeds of more than 10% of the speed of light.

While a solar sail will slow down once the destination star has been reached, a laser sail will not, unless a powerful laser has been built in a previous mission near the target, to perform the braking manoeuvre. A better solution is to have a multistage laser sail, in which the primary sail detaches and acts as a reflector for the secondary sail. The secondary sail can have two parts, enabling the last stage to return to the solar system (Figure 8.5).

Instead of a light beam produced by a laser, a microwave beam generated by a maser could be used. A very interesting concept is for the *Starwisp* probe, a microrobot of just 4 grams, propelled by a sail made of very thin wires which, although having a diameter of about 1 km, would have a mass of just 16 grams. A maser with a power of 10,000 MW would propel the probe to reach a speed of 20% of the speed of light in a few days. It could reach Alpha Centauri in 20 years,

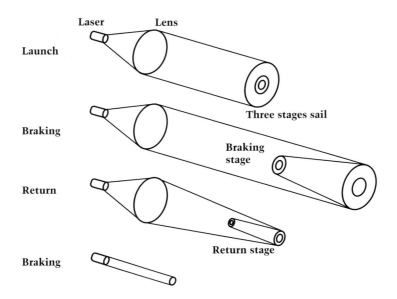

Figure 8.5. Sketch of a multistage laser sail (redrawn from E. Mallove and G. Matloff, *The Starflight Handbook*, John Wiley & Sons, New York, 1989).

where it would broadcast the planned observations back to the Earth using the power of the maser again.

Problems with laser sails are associated with the very large power needed and, above all, with the difficulty of maintaining a very-well-collimated beam stationary in space. The slightest lateral deviation of the beam would cause it to miss the sail, which could not recover itself. There is no chance of controlling the laser beam in such a way that it always points towards the sail. This is because, at a distance of more than a few light hours, the delay in the control loop would make pointing impossible. The task of remaining inside the beam must be entrusted to the sail itself. This problem might be solved by suitably shaping the beam, in such a way that the sail remains near its centre. Scattering of the laser light by interstellar dust is an added problem which might make this scheme impractical.

Interstellar ramjets

The term interstellar ramjet comes from an analogy with atmospheric ramjets. These are jet engines in which air is compressed not by a compressor, as in standard turbojets, but by a converging duct into which the air is forced by the speed of the aircraft. An atmospheric ramjet cannot start from a standstill, but needs a minimum speed to work. An interstellar ramjet is a nuclear fusion rocket which uses the interstellar medium, mainly made up of hydrogen, as fuel. This enters the engine due to the speed of the spacecraft, undergoes a nuclear fusion reaction and is then ejected at very high speed to produce the thrust. The interstellar ramjet cannot work at low speeds, and thus the probe must initially be accelerated by another method.

The real problem here is that the interstellar medium is extremely rarefied, and so the intake of the ramjet must be huge to collect enough hydrogen. As it is absolutely impossible to built a material duct large enough, one proposal has been to capture interstellar ions via electric or magnetic fields, itself a great technological challenge. The interstellar medium is only partially ionised in that part of the galaxy where the solar system is located, and so it has been suggested that lasers could ionise hydrogen atoms, which could then be picked up by the magnetic scoop. For an extremely optimistic value of one billion hydrogen ions per cubic metre, with an intake of at least 160 km diameter, a spaceship with a mass of 1000 tonnes could accelerate indefinitely at an acceleration of 1g, approaching the speed of light in about 1 year. However, being realistic about it, this is not a feasible propulsion device for an interstellar probe, appealing as the concept may be.

Reactionless propulsion

All the propulsion devices considered in previous sections, except for solar and laser sails, are pure reaction engines, i.e. they produce a thrust by accelerating a jet of matter. In a way, sails may be thought of as reaction engines too, as they produce a thrust by reflecting back a stream of photons which hit their surface.

Some reactionless propulsion devices have been proposed; however, they are hypothetical and based on modifications to fundamental physics. Many are meant to allow fast interstellar flights and even to reach relativistic speeds. It has even been suggested that some may break the light barrier to reach superluminal speeds. These concepts will be briefly dealt with in Chapter 10.

One problem is that the rarefied interstellar medium may exert a drag on a spacecraft moving through it at relativistic speeds. Thermo-mechanical problems could threaten the outer shell of the spacecraft. Another significant danger is that of being hit by the tiniest dust speck while crossing the Oort cloud.

PRECURSOR MISSIONS

The machines in which humankind may in the future leave our planetary system may look no more like those considered here than our modern motor cars resemble the 'wind cars' and the other self-propelled vehicles of fifteenth century design studios.

Whilst present technology does not permit the launch of an interstellar probe, some precursor interstellar missions have been proposed and evaluated. As already stated at the end of Chapter 4, the expected scientific return of missions beyond the orbit of Pluto is large. But missions to the edge of interstellar space are even more important from the technological viewpoint, as they will be the *raison d'être* for the development of propulsion devices for very-long-range missions. Crucial tests of the communications, guidance, navigation and other subsystems when operating at such large distances from the Earth will be planned and carried out.

All 'deep space missions' involve a long-term commitment, as the results come in one or more decades after the launch of the space-craft. This means that long-term planning, well beyond the usual horizons of politicians and decision makers, is essential and also that the costs are very high. The costs associated with maintaining a dedicated staff working on the project for decades can exceed that of the probe itself.

Figure 8.6. An artist's impression of the *Pluto Express* probe leaving the Pluto–Charon system, ready for an encounter with an object in the Kuiper belt (NASA drawing).

The first precursor mission could be a flyby of an object in the Kuiper belt. This could be performed quite easily in the first decades of the twenty-first century by the *Pluto Express* spacecraft (Figure 8.6). The next scientific target will be the heliopause and a direct analysis of the interstellar medium. The third type of precursor mission could be one to the focal line of the solar gravitational lens. All the above missions are within the capabilities of conventional propulsion techniques, the main limitations being economic. If the largest non-reusable launchers now available (of the Proton, or Ariane V, class) were to launch into orbit a multistage chemical rocket whose payload is a small probe, they could be accomplished within a reasonable time. The use of innovative propulsion methods is necessary to reduce the flight time and, eventually, launch costs. Precursor missions of this type may best be attempted with solar sail or electric propulsion (e.g., solar electric), like the *Deep Space 1* probe, or even with nuclear electric propulsion.

Far more ambitious targets will be the comets of the Oort cloud. Their large distance from the Earth makes an actual interstellar engine essential. From the viewpoint of the guidance and on-board computers required, an Oort cloud mission is almost as difficult as reaching another star's planetary system. From the Earth it is impossible to locate the object of its final destination. The probe has to identify and find its target by itself and to 'home in' on it, without much help from human controllers on the Earth. When humankind reaches the stage of sending robots to the Oort cloud, we shall indeed be ready for the stars.

MILLIONS OF PLANETS

The true destination of interstellar probes is not the stars themselves but planets orbiting them. The properties of stars are rather well known; the subject of astrophysics informs us about their composition, size, nuclear reactions and evolution, from their birth to their death. If stars were the only objects to be found in the region of the galaxy which surrounds us, we should speak more here about telescopes and other classical astronomical instruments, perhaps located in space or on the Moon, but always in the vicinity of the Earth.

What is really interesting for humankind are the planetary systems which may surround the nearest stars. This is because life as we know it cannot develop in that nuclear hell which is a star, but only on planets or on their satellites. It is also because, if the human species is ever to leave its original planetary system, it will settle on extrasolar planets. The existence of planets around other stars had been predicted when it was discovered that the Sun was a rather ordinary star.

Theories of planetary formation have abounded since Laplace first advanced his theory, in his book *L'Esposition du système du monde*, that the entire solar system evolved from a disc of material from which the Sun, the planets, asteroids and so on, were formed. If planets are a natural outcome of the same process which leads to the formation of stars, the chances are that almost every star has a planetary system.

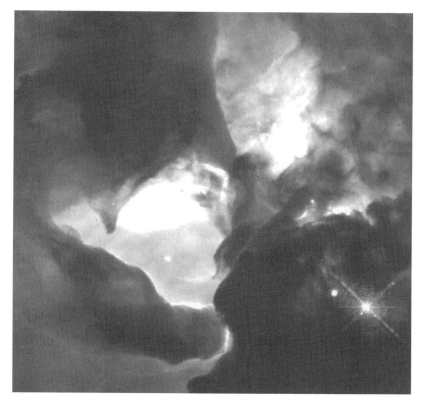

Figure 8.7. A detail of the Lagoon nebula showing gas pillars which have a length of billions of kilometres. New stars are being formed in this region of the galaxy. Image taken by the wide-field planetary camera of the *Hubble Space Telescope*. (NASA photo).

Observations made using the *Hubble Space Telescope* show interstellar gas clouds with stars in the process of formation; one of these 'star nurseries' is the Lagoon Nebula shown in Figure 8.7. When a gas and dust cloud starts collapsing gravitationally to give birth to a star, the conservation of angular momentum causes the cloud to rotate at an ever increasing speed and to flatten out into what is called a protoplanetary disc. At the centre the star (or stars) start(s) condensing, while some of the dust in the disc aggregates to form many objects, with sizes ranging from about 1 to 10 km, termed planetesimals. The planetesimals then collide with each other, producing large rocky planets. The gas giant planets are also believed to have been formed in this way, and their solid cores later attracted some of the gas of the cloud. At the end of the process, most of the mass is concentrated in the central star, while the angular momentum is mostly distributed among the planets, the larger (more distant) ones having most. Protoplanetary discs have been observed around other young stars, and there are good reasons to believe that this process is quite common in the Universe.

If the star is multiple, there is no need for planets to take care of the angular momentum. But the laws of celestial dynamics permit stable planetary orbits to exist even in the case of a double star, provided that the planet's orbit is either very small, around just one of the components, or very large, around both. Alpha Centauri, for instance, a triple star, could have planets orbiting the A or the B components, and they could be planets of the terrestrial type, at distances of the same order as that between the Sun and the Earth. There could also be planets orbiting both components, but they would be on very large orbits, far larger than the orbit of Pluto. The third component, Proxima Centauri, is a dwarf star, and very far from the other two. Even if it has been proven mathematically that stable orbits can exist around multiple stars, it is not at all certain that planets can be formed in such orbits.

It is very difficult to identify a distant planet without any shadow of doubt. Direct observation using optical telescopes is impossible: the very weak reflected light from the planet is completely dominated by light from its parent star. Spectroscopic and other studies are not yet detailed enough to show the existence of planets, at least with presently available instruments. The only way to identify such objects is to study the perturbations which they cause to the orbital motions of other, already known, bodies. There is nothing new in principle here: Neptune, unknown by ancient civilisations, was discovered in 1846 by studying anomalies of the orbits of the other distant planets, which could not be explained by the gravitational attraction due to known bodies. And when the presence of a planet was

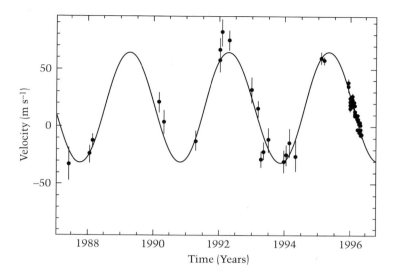

Figure 8.8. Diagram showing the velocity of the star 47 Ursae Majoris measured several times between 1988 and 1996 plotted as a function of time, together with the best fit curve (from R.P. Butler, G.W. Marcy, *The Lick Observatory Planet Search*, Fifth International Conference on Bioastronomy, Capri, July 1996).

predicted in a given position, the planet was actually found as soon as telescopes were pointed in that direction. The discovery of Pluto in 1930 was the outcome of an investigation to explain the perturbations of the orbits of known planets, although it was much more difficult than for Neptune.

To discover planets orbiting a star other than the Sun (i.e. extra-solar planets[7]) is a much more difficult task, as shown by the almost century-long debate on the existence of planets orbiting Barnard's star (see footnote 6 on page 278). Recently, however, the search for extra-solar planets has been performed by measuring the speed of the star, using the Doppler effect, rather than by measuring its position.

The velocity of the star 47 Ursae Majoris, measured several times between 1988 and 1996 is shown in Figure 8.8. A curve fitted

[7] Cole, G.H.A., Exoplanets, astronomy and geophysics, *The Journal of the Royal Astronomical Society*, Vol. 42, pages 1.13–1.17, 2001.

to the experimental results shows that the velocity has a period of 3 years and an amplitude of 48 m/s. Such a pattern suggests that the star has a planet with a mass equal to at least 2.4 times the mass of Jupiter, in a circular orbit around the star with a radius of only 2.1 AU.[8]

With present techniques we can identify only very massive planets orbiting a short distance from the star: if there were a system similar to the solar system, it would be impossible to identify the Earth, and Jupiter would be at the limits of the measurement. The limits of observability with present instruments are plotted in Figure 8.9, together with the mass and the distance from the star of the planets actually discovered by the end of the year 2000. The mass is expressed in multiples of the mass of Jupiter, which is equal to 318 times the mass of the Earth.

Moreover, the possibility of discovering a planetary system depends on the orientation of the plane in which the orbits lie: if this plane is perpendicular to the line of sight, the identification of planets is impossible. The mass of the planet that is measured in this way depends on the angle i between the plane of the orbit and the line of observation: the value derived is not the mass of the planet m but the product $m \times \sin i$. Since the sine of an angle is always smaller than 1, this value is a lower limit to the actual value of m. Starting from this consideration, there are some astronomers who think that planets so 'discovered' are not really planets at all but small stars (brown dwarfs) in much inclined orbits. This objection may be rejected on a statistical basis – it is highly unlikely that all the systems observed have such unfavourable orientations of their orbital planes. Thus, the majority of astronomers consider that extrasolar planets do indeed exist.

Despite all the difficulties involved in the search for extrasolar planets, eight planetary systems were found in less than a year

[8] More recent studies have shown that there are two large planets orbiting 47 Ursae Majoris, with sizes and orbits not too different from those of Jupiter and Saturn. This star could have a planetary system which is similar to our solar system (D. Wright, *Planet Hunters*, www.firstscience.com/site/articles/hunters.asp).

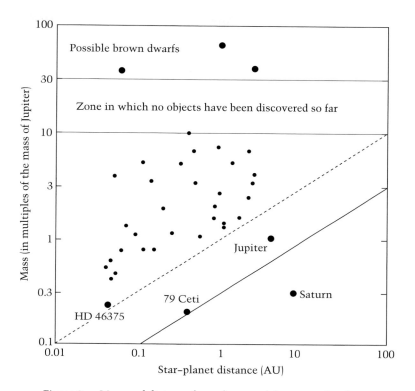

Figure 8.9. Mass and distance from the star of the extrasolar planets discovered by the end of the year 2000. For comparison, Jupiter and Saturn are also shown; the dashed line and the continuous line indicate the limits of observation using different techniques (from the website of the NASA Origins Programme, http://exoplanets.org).

after the first discovery, all of them within a range of 100 light years from the Sun. By the end of the year 2001, some 80 extrasolar planets orbiting stars had been discovered, and in a few cases the extrasolar planetary system was shown to have more than one planet. Two more planets have been discovered around pulsars, while some other discoveries are still not confirmed. Just 6 years after the first confirmed discovery of an extrasolar planet, we may affirm not only that planets exist beyond the solar system, but also that the number of stars having planets is very large. The percentage of stars having a very regular planetary system like ours, in which the orbits of the planets are stable and almost circular, is unknown. Also unknown is whether such systems have a wide variety of bodies – giant planets, made mainly of gas, rocky planets of the terrestrial type, smaller bodies orbiting the larger planets, and a crowd of asteroids and comets. The most promising candidate is 47 Ursae Majoris, a star just 35.9 light years from the Sun. Recent findings indicate that it has two large planets on large and almost circular orbits, a situation which reminds us of the solar system with Jupiter and Saturn. But is it really a planetary system very similar to ours?

One of the stars closest to us (the ninth, in order of distance from the Sun), Epsilon Eridani, a star similar to our Sun even if a little smaller and colder, has a planet with an apparent mass equal to 0.86 the mass of Jupiter, but with a larger semi-axis of its elliptical orbit of 3.3 AU. If it were not for the large eccentricity of the orbit, this could be a planetary system similar to ours. Another, Lalande 21185, has a planet, whose discovery has yet to be been confirmed.

Some of the extrasolar planets recently discovered have masses greater than the mass of Jupiter and their orbits are very elliptical: the existence of smaller planets on stable orbits, above all on 'comfortable' circular orbits, in such systems is questionable. Other planets have masses of the order of that of Jupiter and are very close to their star, closer than the orbit of Mercury around the Sun. This situation is defined as a 'planet of the 51 Pegasi type', from the name of the star around which the first such planet to be discovered orbits. The radius of the orbit of Mercury is 0.39 AU, while that of the orbit of the planet around 51 Pegasi is 0.05 AU. Other planets with even smaller orbits have been found. It is not clear whether massive planets can form on such small orbits; the most common explanation is that the planet formed at a greater distance from the star, and then moved inwards on a spiral trajectory. If this is true, a planet of this type would have completely 'cleaned up' a large zone by colliding with planetesimals or already fully grown planets, to the point that only one large planet exists in the system.

There might be not many star systems containing rocky planets at a distance from the star that is considered suitable for life. On the other hand, large planets close to the star or on elliptical orbits are most readily observed. Therefore, they could be rather rare, appearing to be more common than they actually are.

Planets of the terrestrial type could form as satellites of large gas planets in an orbit close to the star, or form independently and then be captured by a gas planet when close to the star. As an alternative to gravitational capture to become a satellite, a celestial body can be captured at one of the Lagrange points on the orbit of the largest planet. In the solar system this position (on the orbit of Jupiter, 60° ahead of and 60° behind the planet) is occupied by two groups of asteroids, named after Greek and Trojan heroes. For this reason planets or asteroids at the Lagrange points on the orbit of a large planet are said to be 'Trojans'. Even if many planetary systems have no habitable planets of the terrestrial type, they might have habitable satellites or Trojans. They would, however, be extremely difficult to discover from the Earth or from our solar system.

Interferometric techniques, which combine the light captured by two or more telescopes as if they were parts of a single large telescope, may, in the fairly near future, obtain images of extrasolar planets. Through spectroscopic studies, information on the composition of their atmosphere could be obtained and, for example, the presence of oxygen or ozone detected. Such high-resolution interferometers must be located outside the Earth's atmosphere if they operate in the visible. Waiting for the construction of an observatory on the Moon, several projects to build an orbiting interferometer have been proposed. A number of spacecraft, each with a telescope (Figure 8.10), whose position must be controlled with extreme precision, to within the wavelength of light (a fraction of a thousandth of millimetre), is required. This is at the limits of present technological capabilities.

The European Space Agency has selected the IRSI (InfraRed Space Interferometer) *Darwin* mission as a part of its Horizons 2000 programme. This instrument, scheduled for launch in 2015, has six

Figure 8.10. ESA's *Darwin* mission, a large interferometer in space
(Alcatel Space image).

telescopes, all with a mirror of 1.5 m diameter, to be located at a distance of about 1.5 million km from the Earth. In the version shown
in Figure 8.10, the six telescopes are at distances of up to 1 km one
from each other; with a central spacecraft containing the optics that
combines the light coming from the various telescopes. This enables
the light from the central star to be cancelled through destructive
interference, leaving the weak light from the planets to be seen and
analysed.

A NASA mission similar to *Darwin* is the Terrestrial Planet
Finder (TPF) – four telescopes with a diameter of 3.5 m, on four separate spacecraft, plus a fifth platform housing the system to reconstruct
the image. It is to be placed at the Lagrange point L2 of the Sun–Earth
system. Its purpose is to image extrasolar planets of terrestrial size at

distances up to 50 light years away, and to perform spectroscopic studies. The TPF mission belongs to a wider programme, called Origins, whose purpose is to produce a catalogue of the terrestrial planets in that part of the galaxy closest to us and to characterise their atmospheres, with the aim of identifying habitable and possibly inhabited ones. The schedule calls for the launch of TPF in 2011.

A completely different approach to the search for terrestrial planets is to monitor the luminosity of certain stars, in order to detect its decrease when a planet passes in front of the star (partial occultation). To observe the presence of terrestrial planets in this way the star's brightness must be measured with extreme precision (to some parts in 100,000). Such measurements must be performed in space, away from all possible disturbances, with the occultation being identified in a repeatable way. The size of the planet can be assessed from the light reduction, while the time period between two occultations gives information on the orbit and, therefore, the distance of the planet from the star. An advantage of this is that it can be applied to very distant planetary systems. However, if the system is rather close to us, the planet's atmospheric composition might be deduced from the differential absorption of the starlight during occultation.

The probability of a planet being in an orbit such that it passes directly in front of the star is very low. Therefore, many thousands of stars must be continuously monitored to have a good chance of observing just a few events. A NASA mission of this type is *Kepler*, a purpose-designed space telescope. *Kepler* is one of the three candidates for the next phase of the *Discovery* programme and, if selected, it may be launched in 2005. The aim is to keep 100,000 stars like the Sun under observation for a few years, using a telescope with a diameter of about 1 m.

Another proposal for a space mission based on the same principle is the French *COROT*. Actually, owing to the small size of the telescope, the primary goal of *COROT* is not to search for extra-terrestrial planets but to carry out astrophysical studies of stars. But as it should be able to record occultations, its proponents believe

296 BEYOND THE PILLARS OF HERCULES

that it may be able to identify some terrestrial planets during its observations.

Since it is estimated that there are about a hundred billion stars in the Milky Way, countless millions of planets might be expected in this galaxy, our home in the Universe. There may be several only 20 light years from us.

Today, we have no idea about how many (or, indeed, if any) of the extrasolar planets or their satellites are habitable. Neither do we know how many are actually inhabited by some perhaps very primitive, or even advanced, form of life. Further, we do not know how many could be terraformed, nor how many could be made suitable for human life only with great difficulty.

THE FIRST, PROBABLE PROBES

Lalande 21185, a rather small star with a mass of about 35% the mass of the Sun, might have a planet with a mass equal to 90% of that of Jupiter, in an orbit that is slightly larger than Jupiter's orbit. And Lalande 21185 is our fourth nearest neighbour star, just 8.1 light years away. Might this star also have a smaller planet like the Earth, per-haps with oxygen and liquid water? If so, it will become particularly interesting to launch a reconnaissance probe there.

It makes little sense to send a slow probe to such a relatively nearby planetary system. A probe which sends back its findings more than 500 years after it has been launched is a nonsense. Apart from the incredibly high reliability needed for a mission of this kind, it requires that the commitment of keeping in contact with the probe and analysing the radio transmissions which it sends back to the Earth is maintained for many generations. Who will guarantee the huge expenses involved for scientific results whose benefits will come only after 15 generations or more? And will the results be rendered obsolete by the advances in technology made in the meantime?

The minimum requirement for such a mission is a fast inter-stellar propulsion system, to launch a space probe whose velocity v is about $0.1c$. The first results will then appear 50 years after the launch, still a long time. Only a relativistic interstellar flight with, say, a speed of about 0.5c would give results on Alpha Centauri in 13 years.

Advances in miniaturisation will lead to smaller and smaller probes, even though many years will pass before it will be possible to concentrate the payload into 4 grams, as in the case of *Starwisp*. The antenna, to broadcast coded information back to the Earth, and the electrical power system for the transmitter cannot be miniaturised too much, for they must work at such large distances. The probe must operate autonomously, incorporating the most advanced artificial intelligence technologies available just prior to launch.

VON NEUMANN PROBES

Interstellar probes must have an operational autonomy which is close to true intelligence. We do not know how long it will take us to produce a machine on Earth which can pass the 'Turing test', i.e. respond in a way which cannot be distinguished from the response of a human being. Such a machine, provided that it can be built, and then miniaturised, could take command of an interstellar probe and guide it to its target, performing all the tasks entrusted to a human crew. At that point all distinctions between the intelligent machine and the probe would fade, and it would be more appropriate to talk of an intelligent probe.

Once it reaches its target, such a probe could perform an even more important task. It could land on a particular celestial body which it chooses, an asteroid or a planet with a solid surface or a planetary satellite, and start building a copy of itself. An intelligent space probe capable of reproducing itself is a Von Neumann probe. A strategy for space exploration based on these has been proposed by Frank Tipler.[9] A Von Neumann probe could be launched towards a nearby star with a comparatively simple propulsion system. After several hundred years, or even many thousand years, it would reach its destination. The probe would land and start to produce other probes, which would then leave

---

[9] F. J. Tipler, *The Physics of Immortality*, Macmillan, Basingstoke, 1994.

that extrasolar system, heading off for other nearby stars. Once its primary task of continuing the expansion to other solar systems had been fulfilled, the probes would begin their scientific tasks, sending reports back to the Earth. Eventually, most of our galaxy would be settled by these probes. A single intelligent species could even begin to explore the whole Universe using Von Neumann probes. Such intelligent machines might not just explore, but could also reproduce organic life. The question is: how small and lightweight can a Von Neumann probe be? Thanks to rapidly developing nanotechnologies, it may be possible to build a very compact intelligent machine, keeping its mass below 1 kg.

Even when a Von Neumann machine is built, could we be sure that, after many replications of itself, errors would not creep in? After all, this is one of the mechanisms by which evolution creates new living beings. Will a probe programmed on the Earth always perform correctly in the new environments which it will find in other planetary systems? Checking, or even modifying, the programming of the probe by radio from the Earth is possible only for the first few replications. Then the distances in both space and time become so large that everything must be done by the on-board artificial intelligence systems. What might be the outcome of such machines, once they stop behaving exactly as their builders envisaged, owing to random modifications of their genetic code?

Another, more important point has to be addressed. Assuming that such intelligent machines can be built, is it morally acceptable to do so? Should self-replicating machines fill the Universe? That question has caused fierce arguments. Carl Sagan believed the answer to be no. The advisable line for a technological civilisation is that of banning the construction of interstellar Von Neumann machines and strictly limiting their use on its home planet. If the argument of Frank Tipler is accepted, such an invention would jeopardise the whole Universe; the control and the destruction of interstellar Von Neumann machines would then become a task with which all civilised countries – the more technologically advanced, in particular – would in some way have to be involved.

Frank Tipler's answer is equally strong, but positive. If humankind abdicates that role, it will miss all chances of colonising, first, nearby solar systems and then the Universe. Humankind will betray its cosmic duty, and condemn itself to extinction. That reasoning is dictated by fear and ignorance, by a definition by exclusion: what is different from myself is not worthy of existence. A 'person' is defined on the basis of the qualities of the mind and the soul, and not on a particular body form.

By Frank Tipler's reasoning, the dissemination throughout the Universe of Von Neumann machines may be considered as another aspect of that evolutionary process which produced humankind and which may in the future produce other intelligent species to take its place. The ultimate evolutionary task of humans would then be to create intelligent machines, i.e. to move the evolutionary line from beings based on the biology of carbon to beings based on the chemistry of silicon.

PANSPERMIA

The word panspermia indicates those theories concerning the origin of life on Earth being 'seeded' from space, either by natural causes or, in a more fantastic way, by hypothetical intelligent beings. In modern times[10] this idea was put forward by Lord Kelvin, by Hermann von Helmholtz and, in particular, by the Nobel laureate Svante Arrhenius, who probably introduced the term *panspermia*. Arrhenius suggested that micro-organisms can be pushed through space by the pressure of sunlight, to travel not only in the solar system but also in interstellar space. The hypothesis of Arrhenius was revived at the beginning of the 1970s by Fred Hoyle (who died in 2001) and Chandra Wickramasinghe,[11] who considered that they had spectroscopic evidence that much of the interstellar dust contains spores. They thought that comets could carry life not only through the solar

[10] A precursor of this idea was formulated by the Greek philosopher Anaxagoras.
[11] F. Hoyle and C. Wickramasinghe, *Proofs that Life is Cosmic*, Memoirs of the Institute of Fundamental Studies, Sri Lanka, No. 1, 1982; C. Wickramasinghe, *Cosmic Dragons: Life and Death on our Planet*, Souvenir Press, London, 2001.

system, but also in the interstellar medium and through intergalactic space.

The version of panspermia supported by Hoyle and Wickramasinghe has a wider meaning – life had no specific origin, because it has always existed, like matter and energy, for Hoyle was one of the few scientists who did not accept the *Big Bang* theory. In the infinity of time that has preceded us, life diffused through the whole Universe, giving rise to an infinity of intelligent species. Such species had all the time needed to improve the process of panspermia, designing micro-organisms most suitable for space conditions and spreading them into suitable environments. But they went further. They stated that the mechanisms normally considered to be the basis for evolution, gradual mutations of the genomes of the various species, are wrong, or at least insufficient to explain the origin of new species. According to them, evolution proceeds only thanks to the introduction into cells of new genes contained in viruses entering the Earth's atmosphere from space every now and then. If, therefore, panspermia is directed by the will of intelligent beings that preceded us, evolution develops according to a plan, through the insertion of purpose-designed DNA chains. Apart from promoting evolution, viruses also cause illnesses. Hoyle performed epidemiological studies, claiming that illnesses do not propagate by contagious contacts, but are associated with comets entering the Earth's atmosphere.[12] These views have very little following in the scientific community.

Humankind can be an agent of panspermia, by diffusing life through the Universe by sending organic substances or micro-organisms to other planetary systems. Even if this may seem more like the plot for a science fiction tale than an actual possibility, a *Society for the Interstellar Propagation of Life* has been founded. Its promoters start from the idea that life on Earth, as we know it, is something very precious, is perhaps extremely rare and (since there is no proof that life has developed in other places) could well be unique. They think that life is presently much at risk on our planet, due

---

[12] This version of panspermia, sometimes referred to as Cosmic Ancestry, is described in the website www.panspermia.org; it advertises several non-scientific ideas.

both to human activity and to a possible natural event such as a cosmic collision. In this situation they think that it is a basic duty for humankind to do its best to leave the Earth before it is too late to do so.

The means are at hand. Micro-organisms which can survive for a hundred thousand years in the interstellar environment exist, and there is no need to wait for the development of improved propulsion devices able to reach their targets in a shorter time. A probe like *Voyager* could be used, transporting many kilograms of micro-organisms which could be scattered in space at the target extrasolar planetary system. They would then be gravitationally captured and would start the evolution of lifeforms on any planet whose environment encourages that. Or a regatta of thin, light and low cost solar sails could carry an enormous number of micro-organisms to colonise all planets with suitable characteristics, and allow life to take root elsewhere in the Universe. This is the space version of the way in which many plants spread on the Earth, by releasing their seeds in the wind, or the way in which coconut palms colonised all the small islands of the Pacific Ocean through their nuts, carried around randomly by the sea.

An objection to such a programme is that, if micro-organisms sent to other solar systems spread life based on DNA (which developed on the Earth), they might also cause the extinction of other forms of life, different and perhaps already well developed. What for us is sowing life, for others could be seen as encouraging an infection or perpetrating an act of bacteriological warfare. For this reason the promoters of panspermia have been charged with 'DNA imperialism'. Their response is that their aim is to inseminate systems still in their formative stages, systems in which life has not yet started to evolve.

Today panspermia is just an idea, regarded by many as an odd idea. Tomorrow it could be put into practice, by a small group of people, acting on their own. A debate on this theme, which can be listed among other important topics in bioethics, should be started before the possibility becomes feasible. Whilst humans may wish to spread the forms of life from which they evolved to any place which they can reach, as a continuation of that activity which led life to fill every corner of our planet, they should not do so without being fully aware of the consequences of their actions. They should surely not entrust a decision of such importance to a small group of self-appointed individuals.

HUMANS BEYOND THE SOLAR SYSTEM

The problem of the low speed of our rockets travelling such astronomical distances becomes less severe if humans were to embark on journeys lasting several centuries. The most radical solution, even if the least acceptable one, to the problem of settling an extrasolar planetary system with at least one habitable planet is to send frozen human embryos there, together with all the equipment needed to allow them develop, once at their destination, into full human beings. Perhaps more suitable for a horror film than for a science fiction one, this may eventually become feasible. It requires only the most basic interstellar flight technology. The starship can be very small, fairly slow, as the travel time can be measured in centuries, once the problems of conserving the embryos for such a long time have been solved; it does not need the complex life support system which astronauts and cosmonauts need. How these human beings would develop on a new planet, in contact with just machines, is the big unknown. Happily, advances in the field of propulsion will make that solution obsolete from the technical point of view before its implementation stage is reached.

Another radical solution which requires larger speeds, yet still within the realms of slow ($v/c = 0.01$) interstellar flight, is that of a World Ship or a Space Ark. A Space Ark is a space habitat for several hundred people, or even more, which would travel for generations, waiting to reach its destination, an extrasolar planetary system. The inside of a Space Ark would resemble the Earth's environment as far as possible (Figure 8.11). We might consider either wet World Ships, i.e. larger habitats with significant lakes and other water ecosystems, or smaller dry World Ships. The largest conceivable wet ship might have a total mass of the order of 100 million tonnes, and be propelled

Figure 8.11. A Space Ark seen from the garden of another one, as they approach a giant planet (*Ark in Space*, painting by Michael Böhme).

to a speed of 0.02c, with an acceleration of 0.0015g, in about 10 years by an antimatter engine. The habitat itself would have a mass of about 10 million tonnes. After the interstellar cruise, the ark could be slowed down by exploiting the drag of the interstellar gases or by interacting with the stellar wind at the destination star through a magsail.

A starship of this vast size is certainly something still very far in the future; it will have to be extremely reliable. It will need ways to simulate gravity and protect everyone from radiation, in the same way as space habitats must. But the human and social challenges of such an enterprise may be even more formidable than the technical ones. The interactions within a society confined in such an artificial environment for so long are unknown. And establishing the political structure for the necessary decision making is a very delicate issue.

Interstellar exploration and settlement via World Ships is best suited to a scenario in which self-sufficient space habitats have been built in orbit around the Earth or elsewhere within the solar system. Space will by then have become the natural environment of humankind instead of the planets, with the majority of the human population living in space habitats. So why would the inhabitants of a Space Ark, well accustomed to living in space, decide to move their habitat further out into space, in search of a new planet? Is it just human curiosity? Is it to increase the population? Would it not be simpler, and more natural, for people so accustomed to space to build new habitats in space?

Smaller Space Arks may be more feasible. By reducing the mass, the same propulsion device can lead to higher speeds. The optimal trade-off between large and slow or smaller and faster starships must be sought – the optimum solution will surely not favour gargantuan World Ships.

For such a very long journey it could be beneficial for all concerned to slow down the pace of human life. Indeed, nature does something like that for those animals which go into hibernation to survive the winter. The techniques proposed to slow down, or even stop, the pace of human life are many, ranging from hypothermia to various suspended animation practices implemented by nanotechnology. Today, we are beginning to understand the factors which allow many mammals to slow down their metabolism drastically, and to face (without damage) extremely unfavourable environmental conditions such as those encountered during the winter months in many regions of the Earth.[13] It is thus not hibernation brought on by cooling the body (something like deep-freezing the members of the crew before departure, and defrosting them upon arrival), but rather a biochemical process induced by suitable hormones that cause a decrease of the metabolic rate and a lowering of the body temperature. There is no real reason why the process allowing bears and squirrels to survive hostile winter conditions might not also allow humans to face the most hostile environment of all, space.

Hibernation could be very useful not only for long interstellar journeys, but also for travelling in the solar system, particularly if a

[13] T. Kondo, *Approaching Artificial Control of Hibernation*, Third IAA Symposium on Realistic Near-term Advanced Space Missions, Aosta, July 2000.

moderate reduction of the metabolic rate is involved. A hibernating astronaut is not aware of time passing; all the psychological problems related to long journeys through space are removed. Further, the traveller consumes little food, air and water, and may not be subject to the consequences of weightlessness. Moreover, a body in hibernation tolerates higher doses of radiation than a body experiencing normal space conditions, without suffering damage.

These studies are still in their initial phase, but their importance in medicine is such that research work goes on. It is possible that, by the time that propulsion devices for interstellar travel have become available, it will also have become feasible for people to travel for some tens of years and to have their lives extended by tens of years too.

The space travellers will wake up to find that they have been cut off from the cultural evolution of Earth-bound civilisations. They will have to make much effort to catch up with the latest developments, at least in the fields of science and technology. However, the communications between a distant starship and the Earth are so difficult that the adventurers might do better to decide to start a new and totally independent civilisation.

It may not be reasonable to start a journey planned to take several hundred years with the possibility that new technologies, significantly shortening the journey time, would be developed in the meantime. An expedition launched a hundred years later than the first could reach the destination planet earlier. Settlers arriving at a planet after centuries in space, only to find that the planet had already been settled by the descendants of colonists who left the Earth a hundred years later than they had, would receive the ultimate disappointment.

RELATIVISTIC SPEEDS AND HUMAN EXPANSION INTO OUR GALAXY

Nearer to the speed of light, various relativistic effects start to become important. At speeds up to 30% of the speed of light ($0.3c$) relativistic effects are almost negligible. The two most important relativistic effects on spaceflight at higher speeds are the increase of mass, which makes the acceleration of a space vehicle more and more difficult, and time contraction.

Relativistic time contraction (or time dilation) is an actual effect, which has been experimentally verified several times. When nearing the speed of light, the rate of passage of time decreases. Its effect is, in a way, equivalent to biostasis; it could make one-way interstellar trips feasible, particularly if the objective were the colonisation of extrasolar planets. In the case of a round trip, the travellers will find that they have aged less than those who remained on the Earth. This is the well-known twins paradox.[14] The relativistic contraction of time accompanies the increase of mass at speeds close to the speed of light. Of course, the energy needed to accelerate the space vehicle increases very greatly.[15]

From what has been said so far it follows that, if habitable planets do exist at a distance of several light years from the Sun, human beings may well decide to launch colonisation expeditions towards them. However, from considerations of energy, it is unlikely that two-way travel would take place. Nor would commercial or diplomatic relationships exist between such settlements. Maintaining radio contact between two such nearby systems would be difficult, because at least 10 years[16] would pass between sending a message and receiving an answer.

We now formulate an optimistic scenario for human expansion, first to our nearest stellar neighbours, and then to greater distances. We consider that the spatial density of star systems which contain at least one habitable planet is high, with a distance between them, say, of about 10 light years. We also consider that humans are able to

---

[14] A. Hobson, *Physics: Concepts and Connections*, second edition, Chapters 10, 11, Prentice-Hall, New Jersey, 1998.

[15] See Appendix B.

[16] Ten years is the minimum time, in the unlikely case that the average distance between inhabited systems is 5 light years, i.e. that all star systems are inhabited; it is likely that this time would actually be considerably longer, perhaps more like 30 years.

achieve relativistic space travel. In this case the time needed to reach the nearest habitable planet is perhaps 12 years, even if the journey seems to be much shorter to those who travel. After a dozen years from the beginning of the expansion, humans would settle a new planetary system, which may then receive various waves of colonists, depending on many factors, mainly economic (travel cost considerations, economic situation on the Earth, economic prospects of the destination planet, etc.) and sociological influences (demographic pressure, political situation, etc.).

We now assume that the new colony grows in such a way that it can launch a colonisation expedition every two centuries, until it has settled on all the possible targets around it, in all directions. Extrapolating this scenario, human expansion through our galaxy can be visualised as a sphere, centred on the Sun, which expands at a speed of one light year every 20 years.

Because the maximum distance between the solar system and the edge of our galaxy is about 70,000 light years, humanity would have colonised the whole galaxy after about one and a half million years. This is an extremely, unbelievably, short time on the cosmic scale. It is considerably less than a thousandth of the age of our solar system. It is even less than the time taken by our species to settle our entire planet, since the time of the first hominids hunting in the plains of East Africa. But, by then, at least ten billion planets would have been settled at an average of 7,000 new planets each year! This favourable scenario allows an important conclusion to be drawn: the relativistic speed limit for galactic expansion does not prevent humanity from settling on the planets of nearby star systems.

The average expansion speed is determined mainly by the time that a new community takes to become a starting point for a new expansion. If that is a thousand years rather than the 200 years assumed, the expansion rate – though reduced by a factor of five – is still high. The time for an intelligent species to settle a galaxy may approach 10 million years. These figures come mostly from models aimed at demonstrating that extraterrestrial intelligent life is impossible (see Chapter 9), but they can also be used for trying to predict the rate of human expansion through our galaxy.

The greatest problem in this discussion is the definition of a habitable planet. If this term designates a planet on which humans can live, as on the Earth, such planets may be very rare, and it is likely that they may already be inhabited by lifeforms similar to those living on the Earth. The Earth's atmosphere is rich in oxygen, thanks to the work of early creatures, and conditions favourable to life are maintained via a complex self-regulating system in which the biosphere plays an essential role. How common is the presence of some recognizable form of life in our galaxy? We have no answer to this crucial question yet.

The true limits to human expansion beyond the solar system come mostly from the availability of suitable planets to be settled and from our ability to terraform the less hospitable ones. Some of these may even be home to alien lifeforms, perhaps to intelligent forms of life.

The various human communities on different planets will be isolated from one another. Radio contact between them will be rare and not very useful, the information exchanged being at least 10 years old. Commercial activities are not a reasonable possibility.[17]

The situation would be similar to a group of humans becoming cut off from the rest of humankind, as in the case of the Australian aborigines. A better analogy is the settlement of Polynesia: people arrived on various islands at great distances from each other, with technical means (primitive boats) that precluded the possibility of frequent communications between the settlements.[18] And those who travelled had only a very vague knowledge of where the next island was – maybe they were not even sure that it existed at all.

[17] W. Salomon, The economics of interstellar commerce, in *Islands in the Sky*, Wiley, New York, 1996.
[18] B. Finney, *From Sea to Space*, The Macmillan Brown Lectures, Massey University, Hawaii Maritime Centre, 1992.

The most isolated communities on the Earth, even those which remained so for thousands of years, evolved but not to the point of originating a new species. Somatic changes which have occurred in the last few thousand years are so small that a Cro-Magnon man, magically brought into our society today, would only be different from us due to our cultural evolution. But something very different could occur during the process of human expansion beyond the solar system. The strange environments on new planets or their satellites could lead humans to change into distinctly different species. The fact that such relatively small communities would be completely isolated would increase the chances of mutations occurring, as would an environment where the background radiation was stronger than on Earth. The possibility of interventions via genetic engineering could reinforce this trend.

Thus, there would not only be some differences between human beings living on planets and those living in space habitats, but also between different human species occupying different planetary systems. Even if no alien intelligent species exists, it could be humans, as the ethnologist Ben Finney has suggested, who would differentiate sufficiently to produce aliens, but with a common human heritage.

If the contacts between the various colonies were quite rare, such differentiation would be of little significance. After all, who would care much whether the people living in a neighbouring star system which is astronomically close, but very distant on the scale of humans, belong to a slightly different species? While anthropological museums might slowly fill with images of such different human species, to be a curiosity for those interested in anthropology, these different human species would be of little concern in everyday life in the new colonies.

The colonisation of nearby, or even more distant, planetary systems is unlikely to be a realistic means for easing the overpopulation problems of the Earth. It will never be possible for a significant number of human beings to leave our planet to find a better life on some extrasolar system or, for that matter, on some other body in our own solar system. What would be valuable would be for a few members of the human species to establish remote space colonies, thereby enabling the species to perpetuate itself if – or rather when – human life becomes extinct, for whatever reason, on planet Earth.

VIRTUAL TRAVELLERS

The scenarios described in the previous sections are comparatively conventional, the expansion of humankind through the galaxy taking place by moving people out of the solar system to their new colonies. Other radical, and in a way disturbing, scenarios have recently emerged. These bypass the problems linked with large interstellar distances and the huge quantities of energy required to accelerate significant masses to very high speeds.

In previous sections, Von Neumann probes were mentioned, as was the possibility that such machines could, apart from producing copies of themselves, also reproduce organic life once they had arrived at their targets. By using the genetic codes which they would carry onboard, they could also replicate their builders. But this would not be in the sense suggested by the supporters of panspermia, who hold that life colonises the various planetary systems starting on each one of them from micro-organisms and repeating all the stages of evolution. Rather, the various species, including humans, would be reproduced on each planet starting from their genetic codes.

This scenario is not very different from that considered earlier, namely colonisation via frozen human embryos. Here, instead of transporting the embryos, only the genetic information needed to build them is carried, but the final result is quite similar. A 0.1 kg Von Neumann probe could contain all the information required to replicate a town of 10,000 people. All those men and women could be produced starting from many ova, fertilized in artificial uteri, and developed there. After birth they would be brought up by robot nurses. There is, however, an inconsistency in this proposal expressed by those who think that Von Neumann machines are the ultimate product of evolution. Why bother to replicate organic life at all if the true end of evolution is the creation of Von Neumann machines?

Some who think that the human mind is essentially a particular and very complex piece of software running on that hardware, which is our brain, consider that it might be possible to move the same software onto another computer, for example that of an interstellar probe. This statement must be taken literally, and not as an analogy between human and artificial intelligence. Thus we may speak of the emulation of a human being on a particular computer. There is no doubt that it is much easier to transport this man–machine hybrid to interstellar distances than humans in 'their present implementation'. By slowing down the computer's clock, the travel can subjectively become faster.

The authors disagree strongly with this view of an intelligent being. They deny the very possibility of 'implementing a human on a computer', and abhor the ethical implications of such a practice.

There are many intermediate possibilities between a human being as we know him or her now and this human reduced to software running on a computer. There could be a myriad hybrid carbon–silicon beings, partly biological and partly mechanical, the 'cyborgs' of science fiction. They could represent the next stage of our evolution.

However, the very idea of a cyborg is an old idea. The perspectives opened up by nanotechnologies and by genetic engineering point exactly in the opposite direction: many predict that, while the twentieth century was the century of physics, the twenty-first century will be the century of biology. It is thus more likely that we will have biological computers than that silicon computers will be added to living beings. Then the issue will again be an ethical one. Up to what point is the genetic engineering of living beings acceptable? Will it be acceptable to modify our own genetic inheritance?

Another possibility is closer to hand. Rather than physically moving human beings around the galaxy, it might be far simpler to send many probes to record accurately all that is interesting which exists elsewhere, and then to recreate everything here, on the Earth, as virtual reality. After all, an astronaut on the Moon has little direct experience of the lunar environment. He or she can see it only through the small piece of glass of a space suit's helmet, which must also filter out dangerous UV rays, and can touch it only through very thick gloves. The astronaut's other senses do not experience the Moon at all. Today, many people on planet Earth have already experienced the Moon using virtual reality.

The future of space exploration could thus be based on virtual reality. Intelligent machines would scan the Universe, in order to recreate it in the minds of human beings who remain comfortably hidden in their dens on their own planet. This way of exploring the Universe poses a danger in itself. The same technology which can recreate the Universe as virtual reality could also create far more attractive universes, which have nothing to do with the actual Universe. Virtual imagination can take the place of virtual reality.

If the pioneers of the 1950s faced the alternative promises of space flight and nuclear destruction, today the choice is between space exploration and virtual reality.[19] To replace space exploration with its virtual counterpart is a complete betrayal of the space imperative.

[19] B. Parkinson, *Shaping the Flux – the Myth of the Future in Spaceflight*, 48th International Astronautical Congress, Torino, October 1997.

9 Other lives, other civilisations

Are we alone as a (somewhat) intelligent species in the Universe? Although we search for extraterrestrial intelligence using radio telescopes, are we likely to chance upon a distant civilisation trying to make contact with us? Will another civilisation ever listen to the sounds of Earth encoded on the discs aboard the *Voyager* spacecraft?

Would it be wise to advertise our presence in the solar system by deliberately transmitting radio messages? And should we be surprised if an alien form of life were humanoid, i.e. resembling us as human beings? Indeed, what is life?

LIFE IN THE UNIVERSE

'We must admit that in space there are other globes, other races of men and animals':[1] more than two thousand years ago Lucretius had no doubts. Mankind is not alone in the Universe, and many other philosophers and scientists, past and present, would have agreed with him. Clearly a different opinion was held by those who, following Aristotelian orthodoxy, subdivided creation into a sublunary world, inhabited by men, below the sphere of the Moon, and a celestial world, a perfect place where all bodies were perfectly spherical and went around on orbits which were perfectly circular or, at least, made of a combination of perfect circles.

However, 400 years ago Galileo discovered spots on the Sun and saw the mountains and seas of the Moon through his telescope, and Kepler understood that the orbits of the planets were ellipses. The

[1] Lucretius, *De rerum natura*.

idea that creation was divided into two qualitatively different realms become less popular. And, if the planets were not so dissimilar from the Earth, why should they not be inhabited?

For a long time it was a commonly held opinion that the Moon hosted living beings, perhaps intelligent ones. This idea faded when it became clear that the Moon had no atmosphere. For many years intelligent beings were assumed to live on Mars, as considered in Chapter 6. Today, this idea has also been abandoned.

What we know for sure, at present, is that life only exists on one planet. There is no proof yet that life developed on any other celestial body. The only research carried out on another planet, on Mars by the *Viking* probes, although far from conclusive, did not supply any definite evidence for the existence of life there.

There is an indication of extraterrestrial life in the finding of structures, possibly microfossils, in a meteorite (ALH 84001) almost 4 billion years old of Martian origin.[2] This result is still much discussed and, in any event, does not concern present life but life which is now extinct. It has recently been claimed that microfossils have been found in two other meteorites from Mars. The amazing thing is that they are young – they have been dated as about a thousand million years old in one case and only 165 million years old in the other (see Chapter 4). While this could explain why they are more developed than the much earlier one, they also raise questions. Why did life on Mars disappear so recently, when dinosaurs still roamed over the Earth? Or, in spite of this, is life still present on Mars?

Recent discoveries claiming liquid water below a crust of ice on Europa, a satellite of Jupiter, ignite hopes of finding life there. If there is life somewhere on a celestial body, the chances that it is in the solar system are not great; however, the chances increase if the search is extended to nearby (extrasolar) planetary systems, or even throughout our galaxy.

---

[2] S.A. Stern, *Our Worlds: The Magnetism and Thrill of Planetary Exploration*, Cambridge University Press, Cambridge, 1999.

The first issue is to define what life is. When speaking of life in the Universe, it is often specified as 'life as we know it', i.e. life as found on the Earth. If we adopted a broader definition, would we recognise extraterrestrial life?

Bioastronomy, that branch of astronomy studying components for life in other havens, and astrobiology, the branch of biology studying the forms of life which we might encounter in our exploration of space, have recently made progress. Organic compounds and even amino-acids are present in interstellar clouds and comets. Often defined as the 'bricks' from which life is made, they are very common in the Universe, far more common than was believed until a few years ago. One of NASA's main programmes relates to astrobiology and the origins of life.

In spite of intensive research efforts, it is still impossible to give a definite answer to the question: are we alone in the Universe? The answers given in the past have been very different, according to the times and the cultural mores but also the discipline professed by the scientist to whom the question was addressed. Biologists, for instance, have always been much more sceptical, while astronomers and physicists generally tend to think that life is rather widespread in the Universe.

The fundamental questions, defining questions on the existence of life, are threefold. The first is whether the formation of life is a phenomenon which occurred only on our planet or whether some life-forms, however simple and primitive, exist elsewhere in the Universe. The biologists who think that life is due to chance can reach the conclusion that it is an unlikely anomaly, making the Earth unique in the Universe. It has been calculated that the probability of producing a living organism by combining together at random some amino-acids is of the order of 1 in 10^{130}. Comparing this number with the number of stars in the visible Universe, evaluated as 10^{22}, it can be deduced that the probability that living beings exist is unimaginably small. This probability is unbelievably smaller than the proverbial chance of a monkey writing a meaningful book when typing at random on a keyboard. Following this reasoning, we should not exist!

It is sure that, once such a low probability event has occurred, it cannot be repeated. The Nobel laureate Jacques Monod would then be right when asserting that man is alone in the Universe and he is well aware of it.[3]

However, to state that such an unlikely phenomenon occurred by chance is unsatisfactory, the more so today, since we now know that there was only a short time (on a cosmic scale) between the time when life could have started on the Earth and the time when it actually did start. This seems to indicate that the probability that life developed from non-living matter is very high, and so the development of life is – somehow – a direct consequence of some laws of physics and chemistry that cause matter to organise structures of increasing complexity, even living organisms.

Thus, either life on Earth is unique (because it is the product of chance) or life is extremely widespread throughout the Universe (because it is the normal outcome of some physical processes). It is unlikely that a third possibility exists. Some scientists, particularly in the astronomical and space community, agree with the second statement. A few of them are even convinced that the required evidence has already been obtained, thanks to the studies of the meteorites from Mars or other meteorites which are said to contain traces of life.

To answer the question of whether life on the Earth is unique in the negative, the discovery of extinct life would do. What must then be assessed is whether that life originated independently or was carried from one planet to another by meteorites, comets or by other means (see the section on panspermia in Chapter 8).

The second question is whether complex life exists or, as it is often put, whether animal life exists. This term is, perhaps, too influenced by what we know about terrestrial life. It is possible that

---

[3] J. Monod, *Chance and Necessity, An Essay on Natural Philosophy of Modern Biology*, Knopf, Westminster, Maryland, 1971.

complex extraterrestrial life cannot be assimilated to animal, nor to vegetable, life, nor to anything that we know. The answers to this question are therefore much more varied. While the prevailing opinion may be that complex life is common in the Universe, the view synthesised in the so-called *Rare Earth* hypothesis[4] is that, while life is very widespread, complex life is rare. From this point of view, the Earth is unique.

The third question deals with intelligent and conscious life. No general definitions of intelligence or of consciousness have ever been accepted, and we do not know whether the two things are necessarily connected. Perhaps for this reason, the answers here are again very varied. They range from the certainty that intelligence is the normal outcome of the evolution of life to the opinion that intelligence constitutes an anomaly that occurred on the Earth and that soon will be corrected with the disappearance of the human species, perhaps caused by humans themselves.

Despite the many possible answers to these three fundamental questions discussed in lectures, in scientific papers and in popular science books,[5] theoretical science will never be able to determine the existence of extraterrestrial life or of extraterrestrial intelligence. A demonstration of their incompatibility with some basic scientific principles could give us the certainty (for what scientific certainties are worth, since scientific statements must be subject to the possibility of disproof) that they do not exist. On the contrary, a proof of their possibility does not prove that they actually exist.

Experimental science, on the other hand, could prove that extraterrestrial life or extraterrestrial intelligence exist. This could be done by discovering, for instance, that forms of life whose origin differs from that of terrestrial organisms exist, or existed, on Mars. Or it could be done by receiving an artificial radio message from space.

---

[4] P.D. Ward and D. Brownlee, *Rare Earth*, Springer, New York, 2000.
[5] Among many others, see B. Jakowsky, *The Search for Life on Other Planets*, Cambridge University Press, Cambridge, 1998, and R. Shapiro, *Planetary Dreams*, John Wiley & Sons, New York, 1999.

However, the lack of any positive experimental evidence can never be taken as evidence that they do not exist.

SEARCH FOR EXTRATERRESTRIAL INTELLIGENCE

To obtain experimental evidence for the existence of extraterrestrial life there are essentially two lines of research – sending space probes or, later, crewed expeditions to some other planets, or trying to detect the signs of the activities of some alien lifeform from here. Since the chances that life might be found in the solar system are slim, to discover life using space probes we must wait for interstellar exploration missions. In the meantime, the only possibility is to hope that we can observe some signs of life from the Earth or nearby. The assumption that only intelligent life performs activities on a scale large enough to be detected by our instruments has led to the Search for Extraterrestrial Intelligence (SETI) rather than to the search for extraterrestrial life.[6]

If intelligent lifeforms exist, we should be able to identify them from the Earth, either because they do things which betray their presence or because they are intentionally trying to communicate with other species. Up to now, the search for intelligent signals has mainly been concentrated on attempts to receive radio signals, in the belief that this type of transmission is the simplest and most effective means to communicate over interstellar distances. Even if some older precedents exist (Tesla and Marconi thought that they had discovered clues of extraterrestrial radio transmissions), the subject is usually traced back to a famous paper published in 1959 by Giuseppe Cocconi and Philip Morrison in the journal *Nature*, entitled 'Search for interstellar communications'. In that paper the authors demonstrated that radiotelescopes on the Earth could detect possible transmissions broadcast, intentionally or not, as the case may be, by hypothetical civilisations on planets orbiting stars, some tens of light years from us. They concluded that it was possible to pursue experiments on a problem that had previously been dealt with on a more

---

[6] F. Drake and D. Sobel, *Is Anyone out There?* Delacorte Press, New York, 1992.

philosophical level. Radioastronomers found themselves at the front line in a brand new research field which, up to that point, was way beyond their scientific interests.

Although some research is performed in the optical (visible and near infrared) parts of the electromagnetic spectrum, the major effort of recent SETI projects concentrates on scanning the sky with radiotelescopes, searching for regularities which could imply an artificial origin of the signal or even a deliberate attempt to make contact with other species.

The search is based on two different approaches – a selective search on a number of targets, and a search across the spectrum covering the whole sky. In the first case a radiotelescope is aimed at a particular star, chosen for its similarity to our Sun, its age, its distance and the number of planets orbiting it. Not knowing the frequency used by the intelligent beings to broadcast their messages, receivers with a very high number of channels must be used.

Alpha Centauri, for instance, may be an interesting target. Its relative closeness to us means that transmissions broadcast by a hypothetical civilisation on a planet there may be more easily detected here, and because the A component is very similar to our Sun. Unfortunately, it is a triple star; in that case, the planets may not have suitable orbits for the development of life. Barnard's star, almost 6 light years from us, may have a planetary system, but it may be too old for intelligent life on a planet to have survived.

The most interesting stars close to the Sun are Epsilon Eridani, a star only slightly smaller and younger than the Sun, and Tau Ceti, a star similar to the Sun, some 12 light years away. They have been observed in detail. The first caused a false positive result when, at the beginning of the Ozma project,[7] a pulsating signal, very clear and sharp, was received. But the signal was identified only once; after

---

[7] Project Ozma was the first systematic search for extraterrestrial radio signals, carried out in 1960; for details see Y. Terzian and E. Bilson (editors), *Carl Sagan's Universe*, Cambridge University Press, Cambridge, 1997, and N. Prantzos, *Our Cosmic Future: Humanity's Fate in the Universe*, Cambridge University Press, Cambridge, 2000.

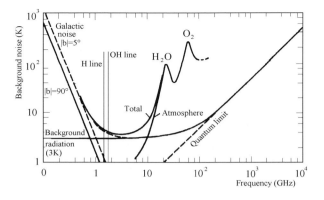

Figure 9.1. Background noise at various radio frequencies. The effect of absorption due to different gases in the atmosphere is reported in terms of noise. The values shown are average values, since the effect depends on many factors, one of which is the direction in which the antenna is pointed.[8]

some time it become clear that the radio signals came from a passing aeroplane!

The second search method analyses signals received by radioastronomers during their long-term studies aimed at other goals, looking for clues of signals of non-natural origin. Signals with wavelengths between 0.18 and 0.21 m (frequency near 1.5 GHz) may be the most suitable for interstellar communication; gases in the interstellar medium do not absorb these signals too much (see Figure 9.1). If such a search is performed in an automated way, its cost can be limited and its coverage of the sky extensive.

No definite result has yet been obtained. So the research must continue, using the best available instruments. However, it will be extremely difficult to obtain a positive result, since we are looking for the proverbial needle in a haystack of cosmic proportions. Neither the origin nor the frequency of the possible signal is known. If only the frequency range of the radio waves is considered, the possible

[8] *Project Cyclops – A Design Study of a System for Detecting Extraterrestrial Intelligent Life*, the SETI League and the SETI Institute, 1996.

channels are many, particularly because the main characteristic of an artificial signal is that it is narrow band, i.e. a signal in which most of the energy is concentrated in a very narrow frequency range. Since the interstellar medium causes frequency dispersion of the radio waves, there is a limit to how narrow the bandwidth can be. Nevertheless, this radio wave region of the spectrum contains billions of possible channels.

What is the preferential frequency for interstellar communications? There is a rather large spectral region where the background noise is relatively low, from slightly less than 1 GHz to about 100 GHz (Figure 9.1). Below 1 GHz the noise produced by our galaxy (mainly due to synchrotron radiation by electrons) becomes the limiting factor. This noise is lowest in a direction perpendicular to the plane of the galaxy ($b = 90°$ in Figure 9.1) and highest at low galactic latitudes, particularly in the direction of the galactic centre. At high frequency, the limiting factor is quantum noise in the receiver. In the optimum frequency range the noise is mostly due to the background radiation produced by the Big Bang, at a temperature of 3 K or, more accurately, 2.7 K.

Since we have no idea of the exact frequency at which the hypothetical extraterrestrials would broadcast, we must use receivers which can receive many channels simultaneously, or record wide bandwidth signals received by radiotelescopes at various locations in such a way as to separate the various channels to study them one by one. In the first case special receivers have been designed, and in the second case computers of enormous power are needed, especially if automatic signal processing is required. As an alternative to a very powerful computer, very many personal computers with relatively low performance can be used. A private organization, the SETI League, supplies computer programs and sets of data to be analysed via the Internet to many thousand volunteers worldwide. Analysis of the signals is performed when the computers have nothing else to do.[9]

[9] Both the programs and data can be downloaded from the website www.seti@home.

The time duration of the transmitted signal is not known and the need to confirm the discovery poses a further problem. Since both the planet from which the transmission comes and the Earth rotate and move through space in a complicated way, the length of time for which the receiving antenna can capture the transmitted beam may be short and the signal might never be received again. This is true for very directional signals transmitted unintentionally towards space, while non-directional transmissions (perhaps similar to our television broadcasts) could be received for much longer times, although necessarily being much weaker. Signals purposely broadcast with the aim of communicating with us should also be receivable for longer times, since those who send them should be aware of the problems of compensating for the motion of the planets and of choosing a not too directional transmitting antenna.

Funding for this research comes from public organisations, foundations and private associations. In the United States SETI initially received good funding from NASA. Then the American Senate cancelled the funds and went to the point of forbidding NASA to deal with this subject; the search continued entirely with private funds, thanks to organisations like the SETI Institute and the SETI League. Currently the veto is less strict, but the research is mostly conducted with private funding. In other nations funds may come from sources similar to those devoted to research in astronomy and radioastronomy.

Many wonder whether it is wise to spend even rather modest sums for research that has such a low probability of success, at least in the short term. But the low probability of success is accompanied by its extreme importance should a positive result be obtained, and there is general agreement in the scientific community to continue.

The confirmation of a positive result would be of momentous significance for humankind.[10] To discover that humanity is not alone in the Universe would have major effects on ideas which we have of

---

[10] J. Heidmann, *Extraterrestrial Intelligence*, Cambridge University Press, Cambridge, 1995.

ourselves and of life in general. An exchange of ideas with a civilisation different from ours may have unimaginable consequences in all fields. But to exchange information with a civilisation 12 or 15 light years away would take a generation. With such a long delay, any conversation would be very stilted.

THE DRAKE EQUATION

Those who search for intelligent lifeforms in the Universe often refer to an equation introduced by the radioastronomer Frank Drake. This expresses the number N of communicative civilisations in our galaxy, whose radio emissions are detectable, in the form[11]

$$N = R \times f_p \times n_e \times f_l \times f_i \times f_c \times L,$$

where R is the number of suitable stars (usually interpreted as solar-type stars) which are formed each year in our galaxy, L is the time for which a civilisation can communicate with us, n_e is the number of planets at a distance from its Sun suitable for life, and f_p, f_l, f_i, and f_c are factors (less than one) which express the cosmic, biological and cultural factors influencing civilisations:

- f_p is the fraction of suitable stars having planets,
- f_l is the fraction of planets, on which life could develop, which actually host lifeforms,
- f_i is the fraction of planets on which life exists, which have intelligent beings, and
- f_c is the fraction of civilisations which have sufficient technology to communicate and which are actually broadcasting to the cosmos.

Although the Drake equation[12] allows us to put the problem on a proper footing, it does not lead to any accurate numerical result for

[11] Drake's equation has been written in several different forms, all essentially equivalent; here it is given in its original form. It had been anticipated, at least qualitatively, by Tsiolkovsky in some of his essays written in the 1930s.
[12] See B. Finney, Section 19.4, in *Keys to Space: An Interdisciplinary Approach to Space Studies*, A. Houston and M. Rycroft (editors), McGraw-Hill, New York, 1999.

N, as most of the coefficients can only be evaluated with a rather large margin of uncertainty.

To have an idea of the results which may be obtained from the Drake equation, we may assume that:

- $R = 10$ stars per year (every year 10 stars suitable for life form in our galaxy which contains some 10^{11} stars – it is a realistic value). Considering only stars like the Sun may be too conservative, as nobody knows whether life might develop at suitable distances from other types of stars. Neither does the equation consider the possibility that two different civilisations, related to different forms of life, might evolve, one after the other, on the same planet.

- n_e may be larger than 1 (particularly if the claims for life on Mars are justified, and if liquid water can be found at larger distances from the Sun than was previously thought, maybe even on Europa). We may assume that $n_e = 3$, an upper bound value for the solar system.

- f_l could be very close to unity (as life on Earth started very quickly, life may appear rapidly wherever it is possible). We may assume that $f_l = 1$, certainly an upper limit, one that might be extremely optimistic. Together with the value of n_e taken, this implies that Mars and Europa are (or, better, have been in the past or will be in the future) inhabited. This implies that all the planets in a zone suitable for life do indeed host life.

- f_p could be much higher than was thought until recently, after the discovery of many extrasolar planets. We may assume that $f_p = 0.1$, an arbitrary but reasonable value, owing to the number of planetary systems discovered (10% of the stars observed have some planets).

- f_i may approach the value of 1, as intelligence could be the logical evolution of all forms of life. We may assume that $f_i = 1$, also an upper limit; it is probably incompatible with the estimates of n_e and of f_l. It implies that all the planets on which life begins will develop intelligent beings.

- f_c depends not only upon a civilisation's technical ability to send radio messages, but also upon the choices made by that intelligent life, considered as their free will, and depending on their interest, fear and xenophobia. After all, we on Earth are capable of doing so but, except for radio and TV transmissions which penetrate the ionosphere and a few intentional, yet highly criticised, messages, we do not have a planned programme aimed at communication with intelligent life elsewhere in the cosmos. Again we may assume that $f_c = 1$, an upper limit. It means that all the civilisations able to communicate are indeed transmitting; to judge from the only known case, its value should be much lower.
- $L = 10,000$ years; such a time for which a civilisation is able to communicate could be overoptimistic.

The most uncertain term is perhaps L, not only concerning doubts about the duration of an intelligent species that has reached the stage of sufficiently advanced technology to allow it to send messages. The simplest reasons for an intelligent species to stop transmitting radio signals may be that it destroys itself, or that it develops technology beyond the stage where it uses radio waves to send information. On the Earth we are already replacing television broadcasting by cable TV, which cannot be received by 'outsiders'. Today, the most powerful transmitting stations are military and scientific radars, but it is possible that, within a relatively short time (measured on the time scale of cosmic phenomena) they may be replaced by devices that work on quite different principles. There are already some signs in this direction: very-long-range communications with space probes in the future may use laser light instead of radio waves. The fact that a laser beam is much better focused than a beam of radio waves makes a laser a much more efficient means of sending information; it also has a much larger bandwidth. But it is also more difficult for those who are searching for signs of intelligent life to identify.

With these (upper limit) values of the parameters, the equation gives the number of civilizations in our galaxy able to contact us as $N = 30,000$. With this value we should expect that the average distance between civilisations which are broadcasting is about 700 light years.

For the sake of playing some more with the numbers, yet without changing the values of R (10) and of f_p (0.1), we could assume that $n_e = 1$ (a more realistic estimate), $f_l = 1$ (again justified by the speed with which life has developed on the Earth), $f_i = 0.01$ (a low value, based on the time needed for evolution on the Earth and on the many catastrophes which could stop that), $f_c = 0.5$ (half of the civilisations able to communicate are indeed transmitting), and $L = 200$ years. These values yield $N = 1$: there is only one civilisation that is transmitting in our galaxy. Because we are not broadcasting any intentional signal, it is not us! Probably its source is thousands of light years away from us and it is below the limit of detectability. If N is small, around 1, we can only give this value a statistical interpretation: it does not mean that there is always some civilisation which is broadcasting messages, since there will be times when there are several transmissions and others when there are none. Besides, we could obtain much lower estimates for N which would lead to the tiniest probability of our receiving messages from another civilisation in our galaxy.

Since it is unlikely that civilisations much more advanced than ours still use radio waves, it might also be useful to search for traces of very-large-scale engineering works performed by extremely advanced civilisations. The physicist Freeman Dyson has suggested that a civilisation exploiting all the energy produced by its central star may build a thin spherical structure, using material obtained from asteroids or other small celestial bodies. Such gigantic objects, usually referred to as Dyson spheres, might have a radius of some Astronomical Units, and include the star and all its inner planets, intercepting all the energy produced inside its surface. The laws of thermodynamics state that a Dyson sphere should send out part of the energy collected in the form of heat at a rather low temperature. That distinctive infrared radiation would be detectable at very large distances. Detailed analysis of the radiation from 54 stars similar to the Sun has been made to look for anomalous emissions in the infrared region of the spectrum. No evidence of Dyson spheres has been obtained. Certainly, 54 stars out of the billions of possible candidates are very few indeed, and the large astroengineering works which a very advanced civilisation might undertake to demonstrate its presence, and its prowess, to others are not limited to Dyson spheres.

The Drake equation has other uses. If the definition of L is revised to be the total duration of a civilisation, instead of the time for which it can send messages, and setting f_c at unity, the equation gives the total number of civilisations currently existing in our galaxy. Or by defining L as the total duration of all forms of life on a planet and setting f_i and f_c at unity, the total number of planetary systems in the Milky Way in which life exists is obtained. It might be between a thousand and a hundred million.

COSMIC AMBASSADORS

Even though we do not broadcast intentional messages into outer space, we send many unintentional transmissions which might allow enterprising extraterrestrials to discover our presence. We have also launched some spacecraft which are now travelling almost in interstellar space. Even though it is very unlikely that another civilisation would stumble on a space probe launched by humankind in the vast expanse of interstellar space, such a possibility cannot be dismissed altogether. For this reason the *Pioneer 10* probe carries a plaque conveying information about its origin and the human beings who built it (Figure 9.2).

The *Voyager* spacecraft carry a disc (Figure 9.3) on which a selection of sounds and images of the Earth is recorded in the hope that,

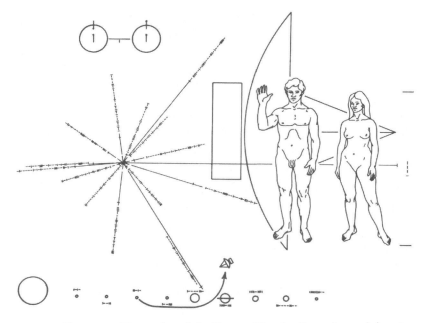

Figure 9.2. Plaque aboard the *Pioneer 10* probe. From the top left, going in a counter clockwise direction, are a sketch of a hydrogen molecule, a map of pulsars with information on the time and place of the probe's launch, a sketch of our solar system, showing the Earth, a drawing of two human figures, and a diagram of the antenna of the probe, to give the scale of the drawing.

if ever they are found, the message could be decoded. The choice of what to include in such a selection, taking into account the small amount of room available on the spacecraft, the need to supply information which could be understood by intelligent beings about which nothing is known, and the requirement to give an unbiased picture of our planet, was a very difficult task. The selection had to be made in a short time, and facing bureaucratic difficulties of various kinds.[13] The disc, a phonograph disc operating at 33 rpm (but recorded to work at 16 rpm), made of gold plated metal (instead of plastic) and protected

[13] See C. Sagan, F.D. Drake, T. Ferris, J. Lomberg and L. Salzman Sagan, *Murmurs of the Earth*, Random House, New York, 1978.

Figure 9.3. Aluminium case containing the gold plated metal disc carried by a *Voyager* spacecraft (NASA photo).

by an aluminium case, has a drawing engraved on its central part describing the procedures to be followed to play the disc.

The selection of black and white, and colour, pictures, and of sounds and music, tries to cover all aspects of our planet, especially of human civilisations. An initial choice was to leave out images related to war, natural disasters or human violence. No religious scene was included, so as not to privilege any religion with respect to the others; the only building in some way related to religion shown is the Taj Mahal, as a monument to love. Religious beliefs come out in some musical pieces, such as those by Bach.

The disc also contains greetings in 54 languages, including some which are no longer spoken, like ancient Sumerian. The thoughts recorded range from colloquial greetings to solemn phrases, such as the Latin *Salvete quicumque estis; bonam ergo vos voluntatem habemus, et pacem per astra ferimus* (greetings to you, whoever you are; we come with goodwill and bring peace among the stars). The Secretary

General of the United Nations, Kurt Waldheim, and the President of the United States, Jimmy Carter, speak, making the *Voyager* the first instrument of galactic diplomacy.

> As the Secretary General of the United Nations, an organisation of 147 member states who represent almost all of the human inhabitants of the planet Earth I send greetings on behalf of the people of our planet. We step out of our solar system into the Universe seeking only peace and friendship, to teach if we are called upon, to be taught if we are fortunate. We know full well that our planet and all its inhabitants are but a small part of the immense Universe that surrounds us and it is with humility and hope that we take this step.

> The message by President Carter ends with these words.

> This is a present from a small distant world, a token of our sounds, our science, our images, our music, our thoughts and our feelings. We are attempting to survive our time, so we may live into yours. We hope someday, having solved the problems we face, to join a community of galactic civilisations. This record represents our hope and our determination, and our goodwill in a vast and awesome universe.

The plaque of *Pioneer* and the discs of the *Voyagers* are thus like bottles containing a message thrown in the immense cosmic ocean in the hope that someone will find them, and learn that an intelligent lifeform existed on the planet Earth, some thousands of years earlier. By that time it is likely that humankind either will have expanded into space or will have blown itself up.

If these messages are found at all, it will probably be by some advanced spacefaring species, with devices developed to find starships adrift in interstellar space. It is extremely unlikely that the probes carrying these messages will eventually enter the inner parts of some distant planetary system.

Just as we send spacecraft which might be found by aliens, humans might find hypothetical alien craft during their exploration of space. If, in the past, an alien exploration team visited the solar system, it is not unreasonable to expect that the visitors left one or more probes here. Certainly, if a probe is left in the solar system with the aim of transmitting data regarding a geologically very active planet like ours for a long time, it is better to locate it in space where it cannot be damaged by volcanic eruptions, earthquakes or other catastrophic events.

The possibility of detecting alien probes in the solar system, whose aim is to study our planet and perhaps even to communicate with us, depends on two factors: the size of the probe and the possibility that it broadcasts transmissions. If the probe is passive, or if its source of energy is exhausted, it may be extremely difficult to identify it. If built with present terrestrial technology such a probe may have a mass of several hundred kilograms and a size of some metres. If we search for an alien probe, we should be looking for something far more advanced technologically. Within one or two decades, micro- and nano-technologies will have progressed to such an extent that a spacecraft may be a single integrated circuit (the so-called probe on a chip), smaller than a credit card. Microprobes will require less power than present probes and, if aliens succeeded in solving the problem of miniaturization of the antenna and the power system, the whole spacecraft could indeed be microscopic. A very advanced technological society might be able to build probes which are invisible to the naked eye, no larger than a speck of interstellar dust ... Could we ever find them?

Alien probes already in the solar system may be 'asleep', ready to be activated by a radio transmission from the Earth. A probe of this type would monitor our transmissions, particularly the Internet, and could have been programmed to talk to us when our message shows that we have reached an adequate technological level.[14]

[14] A. Tough, Interstellar contact: a thousand-year perspective, *Journal of the British Interplanetary Society*, Vol. 52, pages 324–327, 1999.

INTELLIGENT LIFEFORMS

Planet Earth is 4.5 billion years old, and the oldest fossils date back 3.5 (or possibly even 3.8) billion years. Intelligence has existed only for a few million years, if we consider *Homo erectus* as a fully intelligent species, or for slighty more than 100,000 years if we allot to *Homo sapiens sapiens* the status of an intelligent and conscious being. Certainly for no more than one thousandth of the Earth's lifetime has it harboured intelligence.[15] We may therefore think that intelligent lifeforms are extremely rare, but how rare they are depends on how long the species survives. If intelligent species tend to destroy themselves almost as soon as they have reached a technological level such as to permit that, as some pessimists believe, then intelligent life is so rare that there may be no more than one lifeform living at any one time in our galaxy. The same holds for destruction by natural causes, such as an asteroid striking a planet where intelligent life is evolving or the planet crossing a zone of the galaxy where the radiation is especially strong.

The existence of intelligent species in our galaxy and the colonisation of space are, in a sense, antithetical. Since humankind could colonise the whole galaxy in a few million years or, if it did not do so personally, it could entrust the task to Von Neumann machines and, because this statement holds for any technological civilisation, the galaxy should already be swarming with various intelligent species. We should be able to find them everywhere. However, we find no sign of other intelligent species on the Earth or in the solar system. We should conclude either that they do not exist or that the expansion of intelligent species into space is impossible.

The first to realise this contradiction was Tsiolkovsky, who strongly believed both in a large number of intelligent species and in the space imperative. Whilst the Soviet authorities supported the scientific and technological ideas of Tsiolkovsky, a hero of the

---

[15] B. Jakowsky, *The Search for Life on Other Planets*, Cambridge University Press, Cambridge, 1998.

Soviet Union, little was known of his philosophical works until recently. Although unaware of the Russian's essays, the Italian physicist Enrico Fermi pointed out that he had not seen any intelligent extraterrestrials around. And since he strongly believed in the possibility of long-range space travel, this meant that they did not exist! This issue is now known as the Fermi paradox, or better, the Fermi question.

The Fermi question led many supporters of human expansion in space to believe that humans are the only intelligent species in the galaxy, and many SETI researchers to consider that interstellar flight is impossible. On the other hand, Tsiolkovsky believed that such intelligent beings, more advanced than ourselves, refrain from contacting us as it would cause us harm. Humankind might be some sort of experiment which some alien scientists follow from afar without interfering – people behave in just this way when observing animals at a zoo. Some scenarios can be imagined to save the possibility of both extraterrestrial intelligence and interstellar travel and colonisation.

The recent discovery of gamma ray bursters, unimaginably energetic explosions occurring from time to time which destroy all life in a whole galaxy, has led some scientists to believe that it was not until recently (on a cosmic time scale) that life could develop and evolve. Only with a decrease of the rate of such destructive events as the Universe evolves could evolution reach the stage of intelligent life, and we human beings might well be among the first beings to reach that stage. This could be a possible answer to the Fermi question.

It is not at all certain that a species can last for long enough to complete the process of galactic expansion, or that all forms of extraterrestrial intelligence are interested in interstellar travel or colonisation. An extraterrestrial species may have visited the solar system before the Earth was habitable or, for whatever reason, they decided not to stop here.

SETI and astronautics can thus be reconciled – it is only the extreme forms of both that are mutually exclusive, i.e. the statements that intelligent life is very widespread in our galaxy and the strongest forms of the Conscious Life Expansion Principle (CLEP, see Chapter 10).

When considering the Fermi question, we must not forget that some people believe that extraterrestrials are already among us, and that sightings of unidentified flying objects (UFOs) are evidence that their space vehicles sometimes visit the solar system and the Earth. Most sightings of UFOs can be explained by a variety of unusual natural phenomena (e.g., clouds, globular lightning, etc.) or as deliberate hoaxes. But strange ideas on the extraterrestrial origin of UFOs feed a rich market of books and popular articles. Other events connected with UFO sightings are claims of the abductions of humans by extraterrestrial beings.[16]

The hypothesis that somebody else has developed a suitable technology to make an interstellar journey through space cannot be lightly dismissed. The first contact with an extraterrestrial intelligent species may indeed occur on the Earth with the arrival of an alien spaceship. The Fermi question would require exactly this. On the other hand, the authors and most scientists believe that there is no convincing evidence (nor even any serious indications) of the presence of extraterrestrials in the solar system in general or on the Earth in particular.

ET OR ALIEN?

The SETI goes on, with ups and downs as the funding waxes and wanes. Both the Planetary Society and the SETI Institute, private organisations, carry on, ploughing the furrow which was started more than 30 years ago, with improved instruments and better mathematical procedures to analyse the signals. What will happen if we meet with success? How will we behave if and when the fact that someone is trying to establish contact with us is verified?

If the search for extraterrestrial intelligence succeeds, well-defined procedures must be followed. The so called *post-detection*

---

[16] C. Pickover, *The Science of Aliens*, Basic Books, Boulder, Colorado, 1999.

protocols fit within this code of practice. As a first thing, it is important that those who receive an extraterrestrial contact do not divulge the news immediately. They must inform their colleagues and careful cross-checks must be carried out. False positive results have already occurred and a repeat would undermine the credibility of the whole field. The media should be informed only when proof of the existence of extraterrestrial intelligence has been confirmed.

At this point there is another issue to consider. Should we broadcast a response back? If so, who would take the responsibility to speak on behalf of humankind? At present, humanity is not broadcasting messages to the cosmos, telling of our existence.[17] There have been a few such transmissions in the past (see Figure 9.4), which were criticised and then stopped. The difficulty lies with the type of intelligence with whom we might make contact.

Does a general scheme exist, mostly due to the uniformity of the laws of physics, whereby all possible forms of life display a certain uniformity? Is cellular structure necessary for life? Is life necessarily based on carbon, on amino-acids (always the same, with the same spiral direction, i.e. identical chirality), and on DNA? We may imagine beings based on silicon and sulphur instead of on carbon and oxygen. As the variety of silicon compounds is much less than that of carbon compounds, it is likely that life based on silicon is far less rich and, therefore, has a smaller chance of spreading through the Universe.

The questions begin to multiply. Does it make sense to speak of convergent evolution, i.e different species independently evolving similar organs or body forms under similar evolutionary pressures, when two forms of life have evolved completely independently? It is likely that, if intelligent beings evolved on other planets, the configuration of their body is, at least in a general sense, humanoid? Will our, and their, brains allow at least some mutual understanding, or will they be so completely incompatible that nothing is comprehensible? And will that incomprehensibility transform itself into hostility?

[17] What was initially defined as CETI (Communication with ExtraTerrestrial Intelligence) was changed to SETI, which does not involve the exchange of messages.

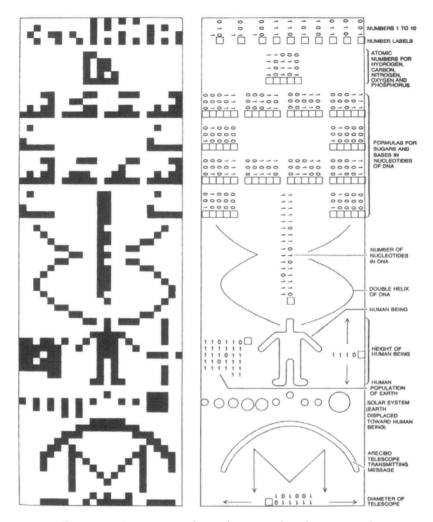

Figure 9.4. A message made up of 73 rows of 23 characters each, broadcast using the Arecibo radiotelescope in Puerto Rico in 1974. The two-dimensional plot of the message (left) is shown together with its explanation (right) (from C. Sagan and F. Drake *The Search for Extraterrestrial Intelligence*. Copyright © 1975 by Scientific American, Inc. All rights reserved.)

HUMANOID CHARACTERISTICS

One much discussed issue is the so-called *predominance of the humanoids*.[18] Must intelligent living creatures resemble, in a general sense, humans? But there is no agreement on what humanoid means: Carl Sagan used the term to indicate a being anatomically equivalent to humans, while the evolutionary biologist G.G. Simpson meant a being with intelligence comparable with human intelligence and any bodily form. Following Carl Sagan, by humanoid we mean an intelligent living being with bilateral symmetry, endowed with two legs and two arms, with a brain in a head (above the bust), in which the main sensor organs are also located, in particular two eyes allowing binocular vision, and a mouth for food.

Such a humanoid configuration is particularly suited to a highly intelligent being. The control system of a standing/walking/running biped is complex; it requires well-developed sensors, a nervous system and a brain. Biped stance and intelligence strengthen each other. If a single 'control centre' evolves, as happened in human beings on the Earth, the requirements for balance control cause an increase of the mass of the brain, and a larger brain uses the two legs in more, and better, ways.

On the Earth evolution is characterised by a gradual reduction of the number of legs that successful creatures have: from the filaments (parapods) of annelida (for instance, the millipedes) to the articulated legs of the arthropods (ten legs for crustacea, e.g. shrimps, eight for arachnids, e.g. spiders, and six for insects). With terrestrial vertebrates the number of legs reduced to four. An animal with a large number of legs, and with a low centre of gravity, does not fall over and finds it easy to walk. A quadruped, particularly if it has a high centre of gravity, must, in each step, go through positions from which it could fall. Therefore, it must coordinate its movements with great precision and have short reaction times.

---

[18] E.J. Coffey, The improbability of behavioural convergence in aliens – behavioural implications of morphology, *Journal of the British Interplanetary Society*, Vol. 38, pages 515–520, 1985.

To go faster an animal must change its gait, no longer walking but running or jumping.[19] Taller animals with higher centres of gravity are generally faster. High speed is important in natural selection, either to run away from predators or to chase food (or to anticipate the actions of other animals in the search for food). Larger animals, with a smaller number of legs, have the advantage here.

The erect position of a biped, with its head above the trunk and not in front of it, is necessitated by the increased weight of the brain. Having the brain in a strong bone box close to the eyes and ears is really advantageous. Possessing arms and hands within the field of view of the eyes is certainly an advantage. Thus, the humanoid configuration seems to be a very practical layout.

A hypothetical example of convergent evolution and of the *predominance of the humanoid form* is the result of a study by palaeontologist Dale Russel. Starting from the hypothesis that the major catastrophe that wiped out the dinosaurs, and therefore led to population of the Earth by mammals, never happened, he simulated the evolution of reptiles toward intelligence. The result was a humanoid 137 cm tall, called *dynoman* (see Figure 9.5): the similarity of the body structure with that of *Homo sapiens* is readily apparent. Palaeontology, like history, is not made with 'ifs' and 65 millions years ago dinosaurs ended their adventure on Earth. *Dynoman* is just an exercise of the imagination, guided by science, but nevertheless a very interesting one.

Whereas the humanoid shape is optimal for an intelligent land animal, it is not suited to aquatic life or to flying. But could aquatic or flying intelligent species evolve? Dolphins started their evolutionary course on dry land, and later adapted to life in the water. In spite of their comparatively big brains, dolphins did not progress further along the road towards intelligence, as they couldnot handle and manipulate objects. It is unlikely that a flying species would develop extreme

---

[19] A.E. Minetti, Invariant aspects of human locomotion in different gravitational environments, *Acta Astronautica*, Vol. 49, pages 191–198, 2001.

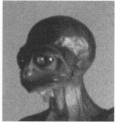

Figure 9.5. Model of a hypothetical humanoid deriving from the evolution of dinosaurs, together with his ancestor, the stenonychosaurus, living from 70 to 80 million years ago in Canada. An enlargement of the head is shown (from T. Dickinson and A. Schaller, *Extraterrestrials – A Field Guide for Earthlings*, Camden House, Camden East, 1994).

intelligence, since the weight of a large brain is certainly an obstacle to flight. However, on planets with low gravity and a dense atmosphere, a flying intelligent species might exist.

What is the optimal size for intelligent beings? There is a minimum size determined by the size of organic molecules which is the same across the whole Universe. Thus, living structures based on the biochemistry we know cannot be too small. If, to reach intelligence, some minimum complexity is needed, intelligent brains have to exceed a certain size and mass. The smallest dimensions of an intelligent being are those able to support, move and feed a brain of sufficient size. The maximum size will depend upon the environment, and particularly the value of the acceleration due to gravity. Larger dimensions can be an advantage, since a big animal can be faster (at least if it is not too big), can defend itself better from predators, see farther, and so on. They are also a drawback, since a big animal needs more energy and must therefore catch more food to eat. And that might help the evolution of intelligence, since the problems created have to be solved. But less time is available for activities not directly related to the search for food. As a general rule, we might expect that intelligent beings which evolved on planets with a lower gravity would be larger. And so we would be surprised to find very small extraterrestrial intelligent beings, for instance as small as an insect or perhaps even as small as a mouse or a rabbit, on a planet such as Mars.

Such reasoning leads us to think that the humanoid form, in size not too different from humans (perhaps from a quarter to four times the size), is the most probable 'layout' for an intelligent extraterrestrial. However, this scenario might be much too anthropomorphic and too influenced by Hollywood. Today, few researchers support the predominance of the humanoid form.

Carl Sagan suggested that living beings with balloons full of gas allowing them to float without ever going down into the denser layers may have evolved in the atmosphere of the giant planets. A hypothetical example is shown in Figure 9.6, a giant intelligent being, perhaps 500 m long, floating in the higher layers of the atmosphere of a giant planet, like Jupiter. If such beings can exist, they could be far more common than the dwellers on rocky planets. We know that extrasolar gaseous giant planets exist, but no rocky planets of the terrestrial type have yet been found. Where – and what – is life, besides that on Earth?

Is this 'floating giant' idea realistic? Probably not for the animal with no hands to manipulate objects, which would be very hard to find in that fluid world. And, if intelligence is accompanied by technology, how could that develop? The inhabitants of the giant planets might reach the same level of intelligence as dolphins.

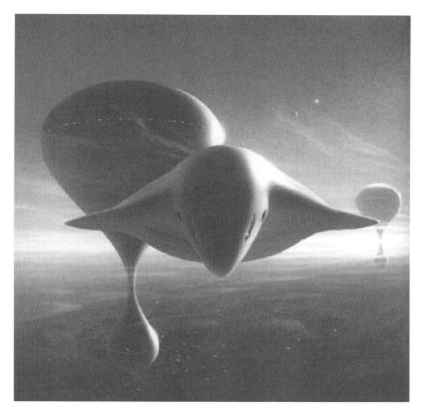

Figure 9.6. Hypothetical intelligent being living in the atmosphere of a giant planet. In the background, one of the cities in which such beings live (from T. Dickinson and A. Schaller, *Extraterrestrials*, Camden House, Camden East, Ontario, 1994).

Humanoids exhibit bilateral symmetry, i.e. they have a plane of symmetry for their external shape, while the organs inside the body may be asymmetrical (the liver on the right, the heart on the left, etc.). This type of plan is typical of most terrestrial animals, from arthropods upwards, following evolutionary lines. And so it is natural for those who designed the being shown in Figure 9.6, with the intention of describing something really alien, to follow this blueprint. Yet bilateral symmetry is not universal in the animal kingdom on the Earth: many sea animals, from starfish to octopuses, have a radial symmetry.

There is no definite 'design' reason for animals living on dry land all to have a bilateral symmetry. It could just be that they all descended from ancestors that had this configuration. If on some planet animals with radial symmetry developed a skeleton and a set of articulated legs to live on the solid surface, it is possible that an intelligent species might have evolved from them. We can imagine configurations as suitable as the humanoid one for hosting intelligence, based on radial symmetry, for instance with three legs, three eyes and three arms.

The debate on the prevalence of humanoids is therefore still open; bearing in mind not that intelligence developed mainly in humanoid species, but that the planets on which humanoids could develop are those where it is more likely that evolution leads to intelligence. Evolution proceeds at random and, when a suitable form is found, works on it with small variations. Surely all possible solutions cannot be explored.

The way in which life started influences these considerations. If life came to the Earth and to other planets from space, as the supporters of panspermia believe, in the form of micro-organisms with their own genetic makeup, we might expect that the beings developed on different planets would bear a closer resemblance to us than they would if their origin were completely independent. The type of intelligence that has evolved may also be similar to ours.

The values which define humanity, for example tolerance, loyalty amongst members in a group and solidarity with fellow human beings, have also been shaped by evolutionary processes. In the long run it is advantageous to stick to such values and further be favoured by natural selection. Similar considerations may also apply to extraterrestrial intelligent beings.

Of course, science fiction writers have imagined everything and the opposite of everything. We may project our hopes or our fears, that the aliens will be useful as wise guides, elder brothers, or friends, helping us to feel less lonely in this vast Universe, or as foes, who compel us to unite and play the role of saviours of humankind.

ET OR ALIEN AGAIN

Is it an ET or an Alien? Or, to mention two successful movies, *Close Encounters* or *Independence Day*? A society so advanced as to be able to communicate and travel across vast distances may, or may not, be advanced morally. It may be that too aggressive an intelligent species would become extinct and that those who survive up to the point of attempting to make contact with other civilisations must be peace-loving. None other than Frank Drake is the source of the pithy thought: 'Bad guys get extinct, good guys survive'.

The chances are that any civilisation which can contact us would be far older and much more advanced than we are. Not only would this make any contact with them important, as we might learn many valuable lessons, but it would in itself be a guarantee against dangers. With scientific and technological advances, the average life-time of the individuals would increase; it is even possible that such advanced intelligent beings would be immortal, in the sense that they could only be killed, either accidentally or deliberately. They might teach their anti-ageing technology to every civilisation with which they are in contact, so that they too might become peace-loving immortal beings.

However, it could be wise 'not to give away our position' with a transmission. After all, most of the relationships between species with which we are familiar, here on our planet, reduce to the prey–predator relationship. Of course, the large distances are here in our favour, but even a hundred light years might not keep us completely safe.

The effort of imagining other living beings, and other intelligent creatures, and different ways of being intelligent is not a sterile exercise. The result may be changes to our views of the Universe and of life – no more an indifferent or even hostile Universe but, on the contrary, a Universe that promotes the organisation of matter, creating those special structures on which life, and then intelligence, are based. Then life can no more be interpreted as an anomaly, but the main road of cosmic evolution.

If life and intelligence turn out to be widespread in the Universe, we cannot avoid extending the respectful concepts that we have developed on our planet to the widest environment. Intelligence and consciousness could be present to different degrees instead of being, as we have the tendency to think, qualities which are either present or absent. If consciousness and intelligence are causally linked with complexity we have to be prepared to meet other beings of any complexity.

Humankind must be ready to recognise as peers all other intelligent and conscious beings that it might find during its expansion in the Universe. Lacking a more general term that synthesises the essence of a conscious and intelligent living being, the meaning of human (man and woman, provided that the gender distinction is applicable) must be extended to include all intelligent species. The discovery of extraterrestrial intelligence will not be an encounter between humans and aliens, but between humans from the planet Earth and other humans living in the depths of space.

10 Towards a galactic civilisation

New means of propulsion giving much higher speeds are essential for the bold enterprise, the dramatic exploration of space. But is it realistic to expect that, in the distant future, we may develop them? Could the speed of light record be broken? Could something 'concrete' become superluminal? This final chapter really pushes the frontiers – some things that are discussed may never come to pass.

Would significant differences develop between human species developing in different environments in space? If so, what might be the consequences? Is it our destiny, our true duty, to go – boldly – forward into space?

BREAKING THE SPEED LIMIT

When journeys over interstellar distances are considered, it is the relativistic speed limit which is the ultimate limit to human freedom of movement through the galaxy. To overcome this limit, we may consider in our imagination a multi-dimensional Universe, in which it is possible to leave our three-dimensional Universe and re-enter our galaxy at another point without actually having to travel between the two points. Then a series of hyperspace journeys, hyperdrives, jump points and many fantastic hyper-manoeuvres can be imagined to allow human beings of all types, good and evil, heroes and cowards, to violate the speed limit imposed by Einstein. In some cases the jump through hyperspace is instantaneous while in others it takes some time, perhaps being proportional to the distance travelled, as if actual movement at a finite speed is happening.

Or perhaps superluminal interstellar flight could be produced by a topological warping due to a curvature of space-time. If objects can be moved through interstellar space, can information also be transferred from one point of the galaxy to another in an equally short time? It is natural to wonder whether these things are just fantasy, a sort of modern mythology, or whether something like this could actually be done, sooner or later.

Experience teaches us that the word 'impossible' must be used with extreme care. There are many cases when it has been used in the past only for later scientific or technological advances to have demonstrated the error of our human ways. And it would be very naive to think that our modern understanding of physics contains all scientific truth so that no further theoretical tenets can be expounded.

At the end of the nineteenth century most physicists were convinced that the 'end of science', in particular physics, had been reached. They felt that all that there was to be discovered had already been discovered. Only some small details had still to be adjusted. Then, when everything seemed settled, some facts that were not well explained by the current theories were reported and small cracks appeared in the edifice of science. To explain these anomalies new theories had to be proposed, and in turn these revolutionised the whole picture. At the start of the twentieth century the failure to measure the motion of the Earth through the cosmic ether and an inability to explain blackbody radiation began a revolution which deeply changed all the sciences. Classical physics made way for relativity and quantum mechanics.

In recent years, some claims which did not find adequate explanations within the framework of currently accepted theories are high-temperature superconductivity, cold fusion (if it really occurs) and the coupling between gravitational and electromagnetic fields. The first is now an undisputed fact, which earned its discoverers the Nobel prize and already finds several practical applications. The other two have not been verified and are still the subjects of heated scientific

Figure 10.1. A very innovative spaceship propelled by a warp drive (NASA drawing, from the BPP programme website).

debates; they may involve misinterpretations of dubious experiments. Both have potentially important consequences for spaceflight, to the point that NASA initiated some research to clarify once and for all whether there is some truth behind the latter. If further experimental tests confirm the existence of these strange phenomena, will their explanations lie within the realms of existing theories?

The Breakthrough Propulsion Physics (BPP) is a NASA programme funded within the Advanced Space Transportation Plan to study innovative concepts which could lead to the necessary revolution in space propulsion (Figure 10.1).[1] At this early stage the work is focused more on its physical and mathematical aspects than on applications. The first goal is to lay out the scientific foundations of what could become a new technology some decades from now – *to make credible progress toward incredible possibilities*. Stress is placed on the low cost of this initial project.

[1] See http://www.grc.nasa.gov/www/bpp.

The three research avenues of the project are conveyed by the words *mass, speed* and *energy*. The directives are to search for:

- New propulsion methods which avoid, or drastically reduce, the need for propellant – in other words, to go beyond present methods of rocket propulsion. This implies finding a new way to move objects, by manipulating inertia, or gravity, or by looking for interactions between matter, fields and space-time.
- A way to reach the maximum possible speed, to reduce flight times by orders of magnitude and, possibly, to circumvent the relativistic speed limit. This implies finding a way to move a space vehicle at speeds close to the maximum speed limit for motions through space, or through the motion of space-time itself.
- New ways to generate huge quantities of energy aboard spacecraft. This goal may have a very broad range of other applications.

All propulsion devices actually in use are based on the thrust which a space vehicle receives when it ejects a propellant at high exhaust velocity, following the basic principles stated by Newton in his second law of motion. The only exception is that of solar (or laser) sails, in which the propulsive force is supplied by the momentum of light radiation hitting a large surface, an application of Newton's third law.

To obtain a thrust without the need to expel mass, new physical principles must be found and then exploited. For example, if negative mass (not to be confused with antimatter) exists, a gravitational repulsive force can be postulated. Or electromagnetic fields might manipulate inertia, gravity or space-time; however, fields many orders of magnitude greater than those possible at present would be required.

Another phenomenon to be investigated is the Casimir effect, in which two closely separated plates are pushed towards each other. This phenomenon may be ascribed to quantum vacuum fluctuations. Finally, some experiments seem to imply that a rotating type II superconductor can affect the gravitational field nearby. If that is true,

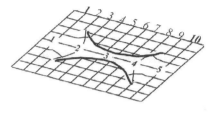

Figure 10.2. A sketch of a two-dimensional wormhole, connecting two points in a plane. The numbers indicate equal distances through normal space and through the wormhole. An actual wormhole is thought to exist in multi-dimensional space-time.

the coupling between gravitational and electromagnetic fields would be several orders of magnitude greater than predicted by the theory of relativity.

The speed limit for both material objects and information comes from the theory of special relativity. General relativity, however, seems to open up the possibility of circumventing this speed limit by altering space-time itself. One way of changing space-time is to create a *wormhole* (Figure 10.2). Wormholes are particularly interesting because, using them, objects could move between two points several light years apart in just a few hours. Because the gravitational effects inside them seem to be very weak, an object could pass through unscathed.

In 1994 Alcubierre[2] suggested that superluminal travel could be achieved by distorting space-time around a spaceship in such a way that the motion does not occur through space at a speed greater than the speed of light, but within a space-time distortion (Figure 10.3). However, this sort of *warp drive* might not be physically attainable.

Other proposals to obtain space drives (see Figure 10.4) based on different principles have been put forward.[3] In particular, the idea that inertia is not an intrinsic property of matter, but a kind of electromagnetic effect due to the zero-point field, opens up new possibilities. This concept, however, rules out the possibility that negative matter exists. Through the manipulation of inertia, masses inside the spaceship

[2] M. Alcubierre, The warp drive: hyper-fast travel within general relativity, *Classical and Quantum Gravity*, Vol. 11, pages L73–L77, 1994.

[3] M. Millis, Challenge to create the space drive, *Journal of Propulsion and Power*, Vol. 13, pages 577–582, 1997.

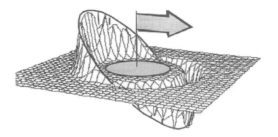

Figure 10.3. Diagram to illustrate the Alcubierre warp drive concept.

are not subjected to inertia forces, thus eliminating one of the greatest problems of high-speed spaceflight. However, this also reduces the forces obtained by ejecting propellant from the spacecraft.

The search for the elusive tachyons, hypothetical subatomic particles which, having negative rest mass, travel faster than light, is

Figure 10.4. Artist's impression of a hypothetical spacecraft with a 'negative energy' induction ring, inspired by recent theories describing how space could be warped with negative energy to produce hyperfast transport (NASA drawing, by artist Les Bossinas).

being pursued. Whilst their existence has been predicted theoretically, they have never been observed. However, even if they exist, they may not interact with ordinary matter; thus, they cannot be seen nor be used to transfer information. Experiments in which photons tunnel across a quantum barrier at a speed greater than the speed of light showed that the speed at which information is carried is less than the speed of light.

Teleportation is one of the tricks of science fiction[4] to move objects, and people, at the speed of light. The basic idea is to dematerialise an object, to extract from it all the information which it contains, to transfer this to another place, and then to organise new matter on the other side of the link to reconstruct the object. Is the original object destroyed? Can the information stored be used to reproduce more than one copy? If the object is a human being, what is the meaning of having copies? Is it like cloning? Is a human being more than a set of atoms organised in a certain way?

In spite of all this, some research on quantum teleportation systems has been undertaken; teleportation of a photon has been achieved.[5] The teleportation of a material object is, however, a completely different matter.

The third goal, the discovering of new means of energy production, is linked with the extraction of energy from quantum energy fluctuations of a vacuum.[6] Surprisingly, this seems not to violate the laws of thermodynamics, but it is not yet known whether vacuum energy exists as predicted. Neither is it clear how much of this energy could be extracted and what the consequences of this would be.

---

[4] The most popular application of teleportation devices, shown in the *Star Trek* television series, was introduced to solve a budgetary problem, namely to avoid the costs associated with a starship landing. That had to be done without computer animations, which at the time the series was produced were very costly special effects. For a thorough discussion of teleportation, including its physical impossibility, see L.M. Krauss, *The Physics of Star Trek*, Basic Books, Boulder, Colorado, 1995.

[5] D. Bouwmeester *et al.*, Experimental quantum teleportation, *Nature*, Vol. 390, pages 575–579, 1997.

[6] D. Cole and H.E. Puthoff, Extracting energy and heat from the vacuum, *Physical Review E*, Vol. 48, pages 1562–1565, 1993.

It is important that such issues are clarified, either to obtain a positive result or to rule them out. If we know that one avenue leads nowhere we have to concentrate our efforts in other directions. So it is possible that humans, sooner or later, could find a way to circumvent the cosmic speed limit. They could perhaps then engineer a propulsion system for interstellar travel at a speed greater than the speed of light. But this may not use wormholes or warp drives; rather it may be based upon a future, unexpected scientific advance.

A GLOBAL VILLAGE ON A GALACTIC SCALE?

How would human life be changed if ever it were possible to fly to distant stars at superluminal speeds? Of course, that would be incredibly costly, in terms of energy and of capital and human investment. However, faster than light (FTL) interstellar journeys will probably never begin or end within a planetary system. Within one stellar system, journeys at a fraction of the speed of light will be necessary. This is because the space within planetary systems will be populated by meteoroids, micrometeorites and dust which could damage the space vehicle.

How long will an interstellar journey take? Several hours will be needed to leave one stellar system and to enter a new one, our colonist's destination. To cross the wormhole will take a few hours more. A journey of a few days is actually incredibly short to move from one star to another, but it is a long one for the traveller concerned.

Is interstellar commerce pure fantasy? What goods would it be worthwhile to transport over such enormous distances? Would it be wise to establish political relationships between different colonies in various planetary systems around the galaxy? Would they inevitably lead to conflicts? It is hard to answer questions such as these.

If humans colonise several planetary systems, each colony will seem to be very much alone. The situation of isolation will have some points in common with that on the Earth in the eighteenth century – distant countries could be reached only after very long journeys which could not be afforded by the vast majority of people.

Information exchange will be fast, but not instantaneous as it is around the world nowadays. The situation will be similar to that existing on Earth before the telegraph and telegram were invented, when the fastest way of transferring information was by mail, or by a courier.

MILLIONS OF HUMAN SPECIES

Superluminal interstellar flight could bring communities which settled the different star systems into contact again, just as the navigation technology of the nineteenth century reunited human communities scattered on the many different Pacific Islands. If the initial interstellar expansion is performed at rather slow speeds, different communities will start a progressive differentiation. Later FTL interstellar flights would mix members of two such different communities together and substantial conflicts might arise. But if, on the contrary, the initial expansion occurs with high speed, or even superluminal, flying, differentiation would not yet have occurred and conflicts could be reduced from the very beginning.

As soon as human beings start living in different environments on different star systems, evolutionary pressures will begin to be strongly felt, and the humans will try to change the environment to adapt it to suit themselves. When human communities start living in permanent settlements where the conditions are very different from those experienced on Earth, a serious ethical problem will be encountered. It is this: at what point should humans be changed to adapt them to their new environment? It would certainly be simpler (and cheaper) to adapt the settlers to live in habitats at a lower pressure than to build structures capable of withstanding normal atmospheric pressure. But would the adaptation of inhabitants of a space station to new conditions mean that they could prevent the arrival of a new contingent from planet Earth?

As we begin to imagine the expansion of the human species into space, our civilisation will have to become culturally prepared. If interstellar flight of any type remains impossible for many centuries, colonisation will proceed in very hostile environments, in space itself or on the Moon, on Mars or on asteroids; the pressure to adapt humans will be great. On the other hand, things may be easier if interstellar flight opens up the colonisation of planets similar to the Earth. The pressure to adapt humans may not be so great if superluminal interstellar flight becomes possible in several centuries.

Other ethical problems, perhaps greater ones, will add to those associated with the differentiation of human species, if contact is established with an intelligent species of extraterrestrial beings. But very little can reliably be said on this issue. There is no point here in unleashing fantasy and imagining many possible scenarios. We should realise that nature can be far more imaginative and creative than humans.

SO LET'S GO!

Over two billion years, life on Earth has continually adapted to the changing environment and, in turn, contributed to changes on our planet – life and the Earth have indeed evolved together. In the last hundred thousand years human beings from the East African plains have explored and settled everywhere on the Earth's surface, hot or cold, dry or wet, hospitable or inhospitable. Civilisation has developed in the last 18 thousand years, since the end of the last ice age.

It is in this context that the authors foresee the continuing evolution of human beings, as human expansion in space begins and gains momentum. What happened in October 1957 opened new perspectives, which deeply changed the lives of human beings. The human species has undoubtedly already benefited from its activities in space. The authors consider that further developments in space activities will benefit humankind further, in many ways.

In the first half of the twentieth century, the idea that human expansion in space is an objective of humankind started to take root. The space imperative cannot be achieved by robotic space exploration or by short missions with crews. It entails the settlement of extraterrestrial celestial bodies and habitats in space. Space stations and space colonies are not just places where some work, which cannot be performed on the Earth, is carried out. They are places where human beings will live and feel at home, develop their own culture and fulfil their dreams. The biosphere of the Earth will expand in all these ways, and in return the Universe (at least the small part of it which can be settled by humans) will be 'humanised'.

At a later date, the human species in space could be substituted by the artificial intelligence machines which it may build. In this case robotic exploration, by Von Neumann probes spreading through our galaxy and into the Universe beyond, could be said to be the ultimate manifestation of the space imperative. The authors are not really sympathetic to this vision. In their view, however, the problem is mitigated by the limits of artificial intelligence, at least by the present types of computers.[7] Perhaps quantum computers could change the situation.

Recently, a new hypothesis has been put forward,[8] the so-called *Conscious-Life Expansion Principle* (often abbreviated as CLEP). An intelligent, self-conscious species evolving on a planet is eventually able to set about space exploration. This enterprise is neither an option nor a casual event in the species' history, but represents an essential way to spread high-level life beyond the place where it developed.

The 'small step' which Neil Armstrong took on the Moon in July 1969, and other steps which future astronauts will take on the surface of Mars or another celestial body, are just a beginning. And there is no doubt that our path will be difficult and hazardous. The road to space seems to become harder, rather than easier, as we go on, and harder than was expected two or three decades ago.

Great technological advances are needed, both widening the range of human activities in space and greatly reducing the cost of

---

[7] See R. Penrose, *The Emperor's New Mind*, Oxford University Press, Oxford, 1989.
[8] G. Vulpetti, *On the Viability of Interstellar Flight*, 49th International Astronautical Congress, Melbourne, September–October 1998; G. Vulpetti, Problems and perspectives in interstellar exploration, *Journal of the British Interplanetary Society*, Vol. 52, pages 307–323, 1999.

access to space. The investments that will be required are huge, but the returns would be incalculable, not only in terms of human expansion into space but also for applications for humanity on Earth.

If the Conscious Life Expansion Principle has some basis, the range of human activities carried out in space will become much larger than those required for exploring our solar system. Interstellar distances are so large that our present technology is inadequate to deal with them – here, scientific advances are required.

Although some consider that interstellar journeys will never be attempted, mainly on the grounds that they will always be so costly that there will never be a valid enough reason to embark on them, we predict that slow interstellar travel will become an actual possibility in the future, in a much more developed human civilisation.

But we cannot say when the amounts of energy needed for this enterprise will become available at an acceptable cost. Nevertheless, research on new sources of energy is now being given high priority by the present needs of humankind on its own planet. This leads us to guess that a couple of centuries will be sufficient to start human interstellar expansion to colonise the planetary system of a nearby star.

The possibility of relativistic interstellar flight is more uncertain, while superluminal travel may always be impossible for theoretical reasons. Only FTL travel will allow two-way journeys between the planetary systems of different stars. That will be a true interstellar civilisation, not just the sum of isolated independent cultures. Whether it will be a better understanding of theoretical physics or the emergence of new theories which will allow that is still unknown, and could well remain so for many centuries.

Technological hurdles linked with the very large distances to be travelled, will remain, as will psychological issues due to the inherent nature of human beings. And human beings will find it very challenging to live in any environment so different from their birth place.

We have no doubt that in the future humans will live in artificial environments on the Moon or on Mars, but will they be radically more artificial than that of an air conditioned hotel room here on Earth? Of course they will have to deal with a lower external pressure and screen against harmful radiation. But a human can no more live in the conditions existing outside an airliner than he can live in space outside a space station, even though flying through the atmosphere at heights near 10 km has become such a commonplace experience for millions of humans that they do not even realise what lies just outside the thin protective shell which is the pressurised fuselage.

Human life in a space habitat or on another celestial body will be made possible by a combination of the twin strategies of creating a 'shirtsleeves' environment in the midst of the most hostile of surroundings and of transforming the planets on which humans will settle. Micro-organisms terraformed the Earth in ancient times. Now it is time for humans to proceed along the same track, transforming the environment of the planets which they will settle and simultaneously transforming themselves into space creatures (Figure 10.5).

Space activities can help us to solve problems on Earth such as pollution and lack of resources. Thus, the space option becomes the high road to solve some of our present difficulties, a road to decide to take without delay.

The developments outlined in this book imply a vision, a dedication to the task and an ability to plan, as needed for all momentous human enterprises but which in recent years seem to be fading away. Today, humans, whilst being aware of the power which modern technology puts into their hands, often seem afraid to apply it in useful new ways. They are abdicating that role of the active promoters of change which human beings have always played.

Today in some quarters anti-technological attitudes prevail: there is a general mistrust of science and technology and a longing for a past in which humans and nature were reconciled, but which never existed. Regrettably, scientific and mathematical illiteracy is

Figure 10.5. Man as a space creature; the astronaut Bruce McCandless performing Extra Vehicular Activity, wearing the Manned Manoeuvring Unit (MMU) on his back (NASA photo taken from the Space Shuttle *Challenger* on February 9, 1984).

widespread among people who have a sound general knowledge and, which is worse, among politicians entrusted to make responsible decisions on behalf of us all. Together with the risk-averse culture and stifling bureaucratic attitudes of many, all these failings make new enterprises increasingly difficult to start.

The success of such a bold enterprise cannot be taken for granted. We do not lack historical precedents here. Ben Finney[9] notes that, a short time before Europeans started the age of discoveries at the end of the fifteenth century, China was well on the same way. The

[9] B.R. Finney, The prince and the eunuch, in *Interstellar Migration and the Human Experience*, University of California Press, Berkeley, California, 1985.

conservatism, bureaucratic attitudes and closed minds of the ruling classes wrecked the efforts of enlightened emperors and admirals in a few years, even to the point of destroying the large fleets which started this stunning work of exploration. China then dissociated itself from world activities. It became a recluse for a long period, from which it seems to have emerged only in recent times.

Humanity must recover its ability to think and act constructively on a grander scale. Our civilisation will then be able to spread its wings, to broaden its horizons, first to include space habitats, a base on our satellite, the Moon, and then on Mars, and then onto other celestial bodies at greater and greater distances.

Public bodies are finding it increasingly hard to maintain long-term commitments. Thus, private organisations must be given the opportunity of guiding this process. The role which governments will play is to state 'the rules of the game', to catalyse action through public–private partnerships, and to offer incentives in order to maximise the benefits for all.

As a final consideration, we must not delude ourselves into thinking that technological progress automatically leads to our moral improvement. When our frontiers enlarge, and include new horizons far from our original planet, humans will carry with them all their good and evil. For this reason our ethics have to develop too, particularly if during this expansion in space we come into contact with unexpected lifeforms or new forms of intelligence.

So let us now rise from the planet where humankind started its adventure to face our future challenges and responsibilities as a space-faring species.

Appendix A   Distances in the solar system and beyond

If space is indeed the final frontier of humankind, its size, and its scale, are such as to defy imagination. To say that space is infinite is scientifically incorrect, but true from the point of view of our human perception.

The Earth–Moon distance, which was covered some 30 years ago by a few men in their two-ways journeys, is 384,000 km. As the radius of the Earth is 6,378 km, the Moon is about 60 Earth radii from our planet. Light takes slightly more than a second to go from the Earth to the Moon, while light coming from the Sun takes 8 minutes to reach us. The Earth is a member of the solar system, whose diameter is so large that light takes 22 hours to cross it.

The Système Internationale (SI) unit for length is the metre (m), originally defined in 1791 as 'the 1/10,000,000 part of the arc of meridian from the North Pole to the equator passing through Paris', i.e. about the 1/40,000,000 part of the Earth's equator. It is now defined as '1,650,763.73 times the wavelength in vacuum of the red-orange light emitted by krypton 86 due to the natural transition from the level $2p_{10}$ to the level $5d_5$, when the lamp is at the temperature of the triple point of nitrogen'. One kilometre (km, with the k always written with a small letter), is one thousand metres. Following the SI rule for the formation of multiples, there is then the Mm (Megametre, one million metres or 1000 km), the Gm (Gigametre, 10^9 m or a million kilometres), the Tm (Terametre, 10^{12} m or a billion kilometres), the Pm (Petametre, 10^{15} m or a million million kilometres) and lastly the Em (Esametre, 10^{18} m, or one million billion kilometres).

The smallest unit used in astronomy is the Astronomical Unit, AU, equal to the average Sun–Earth distance, 149.47 million km (149.47 Gm).

The average radius of the orbit of Pluto around the Sun is 5.9 billion kilometres, or 39.6 AU, but the orbit of Pluto is a rather elongated ellipse. The outer limit of the solar system is often taken as the orbit of Neptune, which has an average diameter of 4.49 billion kilometres (4.49 Tm), and which is close to 30 AU. Actually, the solar system is much bigger than both of these distances – it reaches out to the limits of the Oort cloud of comets. An estimate of the radius of the solar system to include the Oort cloud is between 60,000 and 100,000 AU, i.e. at least 2,000 times the radius of Neptune's orbit.

The unit used for interstellar distances is the light year, i.e. the distance which light travels in one year. With the speed of light in a vacuum being about 300,000 km/s (more precisely 299,792.458 km/s), it is equivalent to 9,461 billion kilometres (9.461 Pm) or to about 63,300 AU.

Another unit often used when measuring astronomical distances is the parsec (pc), a contracted form for one-second-parallax. This is defined as the distance from which the diameter of the Earth's orbit about the Sun is seen to make an angle of one arc second, i.e. one second of a degree (1/3,600 deg). This is equivalent to the distance of a hypothetical star whose parallax due to the motion of the Earth around the Sun is equal to a second of one degree. The definition of the parsec comes from the practice of measuring the distances to stars through parallax measurements. One pc is equal to 3.26 light years, i.e. 206,265 AU or 30,830 billion kilometres (30.83 Pm). Proxima Centauri, the star nearest to us, is at a distance of 4.3 light years, or 1.32 parsecs (pc), or about 42,000 billion kilometres.

The estimated diameter of our galaxy, the Milky Way (Figure A.1), is about 100,000 light years, i.e. about 30,000 pc (or about 1,000 Em). The distance of the Sun from the galactic centre is about 30,000 light years, or 10,000 pc.

There are 75 known star systems, containing 105 stars, which lie within a sphere with a radius of 21 light years centred on the Sun. If the density of the stars is everywhere the same, a reasonable assumption for the galactic disc, the number of stars in a sphere centred on the

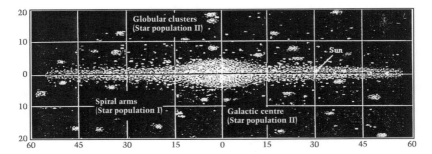

Figure A.1. A cross section of our galaxy, with the position of the Sun being marked (the scales are in thousands of light years).

Sun grows as the cube of the radius. There should be about 1,000 stars in a sphere of 40 light years radius, and about 10,000 in one having a radius of 80 light years.

The two nearest galaxies, the large and small Magellanic clouds, two irregular satellite galaxies of our galaxy, are at a distance of about 160,000 light years, i.e. 50,000 pc. The nearest spiral galaxy, the Andromeda galaxy, at a distance of 2 million light years, belongs together with the Milky Way and many other galaxies to the so-called local group of galaxies.

The diameter of the observable Universe is about 15 billion light years. Such a distance is equal to 1.5×10^{23} (15 followed by 22 zeros) km.

In the SI system of units, besides metre (m) for length, the main fundamental units are the kilogram (kg) for mass, and the second (s) for time.

Appendix B   **The basics of astrodynamics**

The term *astrodynamics* indicates the application of mechanics to the motions of spacecraft. The laws governing the motion of both natural celestial bodies and spacecraft are the same. However, in the second case, more-complex situations arise, such as the effects of engine thrust and of trajectory corrections. Comets are also subjected to a thrust, similar to that acting on a space vehicle, due to their being heated by the Sun and losing mass.

MOTION OF PROJECTILES IN A GRAVITATIONAL FIELD

Since Galileo's time, it has been well known that projectiles move on parabolic trajectories. This result is, however, approximate; aerodynamic drag has been neglected and the gravitational field is taken to be constant, as would occur for an infinite, flat Earth. The trajectory is actually an arc of an ellipse but, if the range and the maximum height are very small compared with the radius of the Earth, the approximation is so good as to be applicable to any practical case.

The gravitational force F in Newtons (N) which acts between two bodies of mass m_1 (kg), the Earth, for instance, and m_2 (kg), for example a space vehicle, with d being the distance (m) between their centres, is

$$F = G\frac{m_1 m_2}{d^2} \tag{B.1}$$

where G is the gravitational constant, whose value is 6.67×10^{-11} (N m^2)/kg^2.

Consider a body of mass m located near a spherical celestial body whose mass is M. The latter behaves as a point mass located at its

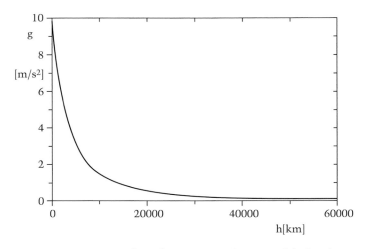

Figure B.1. Gravitational acceleration g as a function of the height
$h = d - R$ above the surface of the Earth, in kilometres.

centre and the force, obtained through Equation (B.1), can be computed
as the product of the mass m and the gravitational acceleration

$$g = G\frac{M}{d^2} \qquad (B.2)$$

where d is the distance between the body of mass m and the centre of
the spherical celestial body. If, instead of the distance d, the radius R
is introduced into Equation (B.2), the gravitational acceleration g_0 on
the surface of the celestial body is obtained. For the Earth, its value is
$g_0 = 9.81$ (or better, 9.806056 at 45° latitude) m/s^2.

The gravitational acceleration g is plotted as a function of the
height above the surface of the Earth in Figure B.1; g decreases with
increasing altitude. This reduction of g is negligible at altitudes of
'aeronautical' interest, i.e. at up to 20,000 m (20 km), and is still small
in the case of a satellite in LEO (Low Earth Orbit). At a height of about
13,000 km, just more than two Earth radii, g is reduced to one tenth
of its value on the Earth surface.

KEPLERIAN TRAJECTORIES

The orbital velocity v_s of a satellite of mass m in a circular orbit of
radius d around a spherical body of mass M is found by putting the

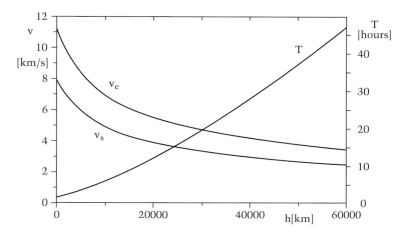

Figure B.2. Variation of the orbital velocity v_s, escape velocity v_e, and orbital period T for circular orbits at height $h = d - R$ above the surface of the Earth.

centripetal force

$$F_c = m\frac{v_s^2}{d} \tag{B.3}$$

equal to the gravitational attraction

$$F_g = G\frac{Mm}{d^2} \tag{B.4}$$

Here the magnitudes of vector forces, **F**, are shown. Then

$$\frac{v_s^2}{d} = \frac{GM}{d^2} \tag{B.5}$$

so that

$$v_s = \sqrt{\frac{GM}{d}} \tag{B.6}$$

The orbital velocity of a satellite is greater for low Earth orbits, and decreases with increasing altitude;[1] $v_s = 7.91$ km/s or 28,460 km/h in LEO (Figure B.2). The speed needed to put a satellite into orbit is greater than that given by Equation (B.6), because the speed to reach the required altitude must also be considered.

[1] What can be learnt about Earth's atmosphere and gravity from observations of satellite orbits is elegantly discussed in D. King-Hele, *A Tapestry of Orbits*, Cambridge University Press, Cambridge, 1992.

If the satellite's speed is lower than that needed for a circular orbit, an elliptical orbit lying inside the circular orbit is obtained. Or, if it is larger, the elliptical orbit lies outside the circular orbit. As Kepler found out, the centre of the body around which the satellite orbits is at one of the two foci of the elliptical orbit.

The eccentricity of the elliptical orbit grows with increasing speed. Thus, the other focus lies further and further away until it is at infinity for a well-defined value of the speed. An ellipse with one focus at infinity is a parabola, and the speed at which a parabolic trajectory is obtained is the escape velocity

$$v_e = \sqrt{\frac{2GM}{d}} = \sqrt{2}v_s \tag{B.7}$$

The escape velocity is $\sqrt{2}$ times the orbital velocity. The escape velocity also decreases with increasing altitude. For the Earth $v_e =$ 11.18 km/s or 40,250 km/h, if the spacecraft is launched from the surface of our planet (Figure B.2).

A space vehicle leaving a celestial body at the escape velocity slows down while travelling along the parabolic trajectory. Its velocity reduces to zero when the distance becomes infinite. In practice, this happens when the distance is large enough that the attraction of the celestial body becomes negligible.

Astrodynamics generally deals with the so called 'two-body problem', i.e. only two bodies interacting gravitationally are considered; each celestial body is assigned a *sphere of influence*. Thus, a space vehicle leaving the Earth and directed at Mars is considered to be in the sphere of influence of the Earth (so that the trajectory is computed as if the Earth and the spacecraft were the only two objects in the Universe) until the attraction of the Earth is no longer much greater than that of the Sun and/or of other planetary bodies. When the attraction of the Earth has become so weak that it is smaller than that of the Sun, the space vehicle is considered to be beyond the sphere of influence of the Earth and within that of the Sun. Then, when it gets close to Mars, it enters the sphere of influence of that planet.

If the space vehicle's velocity is greater than the escape velocity, the trajectory becomes a hyperbola and the speed, although decreasing with increasing distance, never goes to zero. At infinity, in practice when leaving the sphere of influence, the speed reduces to the hyperbolic excess speed

$$v_{\text{hyp}} = \sqrt{v_0^2 - v_e^2} \qquad\qquad (B.8)$$

where v_0 is the initial velocity.

The possible trajectory which a space vehicle may follow in the gravitational field of a spherical celestial body is an ellipse, parabola or hyperbola; these are all Keplerian trajectories. The simplest approach to the design of a space trajectory is to consider it to be made up of arcs of Keplerian trajectories, each of which is obtained considering only the space vehicle and one celestial body, in whose sphere of influence it is.

Let us consider, for example, the trajectory of a space vehicle going from the Earth to Mars. Both planets are on elliptical trajectories around the Sun with low eccentricity and, for a simpler calculation, we may assume that they are on circular trajectories. The most efficient trajectory, from the energy viewpoint, between two circular orbits is half an ellipse which is tangent to the two circles (the Hohmann transfer orbit[2]). In the case of Mars (Figure B.3), to get into that transfer trajectory the speed at point A must have a value of 32.6 km/s, while the speed of the Earth in its orbit around the Sun is 29.8 km/s. To reach Mars, the speed which the space vehicle already has due to its having been launched from the Earth moving around the Sun must be increased by $32.6 - 29.8 = 2.8$ km/s. The space vehicle must exit the sphere of influence of the Earth with a hyperbolic excess speed of 2.8 km/s. If it is to be launched from the Earth's surface, its initial speed, following Equation (B.7), must be $\sqrt{11.18^2 + 2.8^2} = 11.5$ km/s; this is only slightly greater than the escape velocity.

---

[2] Further details are given in R.X. Meyer, *Elements of Space Technology*, Academic Press, London, 1999.

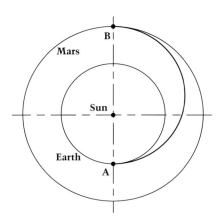

Figure B.3. Minimum energy transfer trajectory between the orbits of the Earth and Mars (the Hohmann transfer orbit).

Instead of launching the spacecraft from the Earth with a speed of 11.5 km/s, it could be launched with the escape velocity, and then, when outside the Earth's sphere of influence, a speed increment of 2.8 km/s would enable it to enter the trans-Mars trajectory. The rocket engines would then impart a total velocity increment of $11.18 + 2.8 = 13.98$ km/s, greater than the value of 11.5 km/s obtained earlier. This is an example of a general rule: from an energy viewpoint it is preferable to apply the propulsive thrust as close to the planet as possible, i.e. as deep in the gravity well of a planet as possible.

Often this is not done, for practical reasons. It is generally convenient to launch the space probe into a low Earth orbit, and then to inject it into a planetary transfer orbit. This is a safer practice, waiting for the proper launch window and making all the last minute checks before the final burn. And a low Earth orbit is well within the gravitational well of planet Earth to exploit most of the advantage mentioned above.

PERTURBATIONS TO KEPLERIAN TRAJECTORIES

The approach explained in the previous section is not adequate to compute space trajectories with the greatest precision, as astronomers realised when they tried to predict eclipses of the Sun and Moon. Celestial bodies are not exactly spherical, and this means that the attracting body cannot be considered as a point mass in its centre – no actual orbit will be exactly elliptical.

The assumption that in each part of the spaceflight only one celestial body interacts gravitationally with the spacecraft is an oversimplification. Many celestial bodies simultaneously attract the spacecraft and the gravitational field in the solar system is complicated. The gravitational fields of the Sun and Jupiter must be taken into account, while the effects of the other planets may be neglected. Further, the orbits of the various planets do not lie in exactly the same plane, Pluto's orbital plane being well off the ecliptic plane. Other effects, including relativistic effects for the orbit of Mercury due to its relative closeness to the Sun, complicate this picture.

The traditional way of accounting for these effects is to compute the spacecraft trajectory using the simplified two-body approach, and then to introduce the required corrections in the form of 'perturbations' due to the other bodies and to non-spherical shapes. The pressure of the Sun's light and, for low Earth orbits, the frictional drag of the upper atmosphere are also accounted for as perturbations.

The more modern approach involves numerical integration of the equations of motion of the general multibody problem with time. With powerful computers, this approach allows us to obtain incredible accuracy – for instance, when the *Voyager 2* probe approached Neptune, after a journey of 8 billion kilometres, it was off course by just 30 km. Very precise tracking makes this precision possible; for the *Voyager 2*, the uncertainty in its position was only some tens of metres.

SPEED INCREMENTS

The trajectory of a spacecraft is made up of arcs of ellipses, parabolas or hyperbolas, at least for the two-body assumption, if the space vehicle proceeds by its own inertia, without any thrust being applied to it. When chemical rockets, which produce high thrusts for short times, are used the trajectory is considered as being made up of arcs of these curves; the short impulsive phases between them are usually assumed to be instantaneous, each being characterised by a speed increment Δv. Each manoeuvre is characterised by a certain Δv, and the total speed increment Δv_t needed to perform a particular mission is the sum of all these increments.

For a lunar mission, performed with a stay in a parking orbit around the Earth, and another parking orbit around the Moon, the following values of Δv can be computed as:

- launch to LEO $\Delta v = 7.91$ km/s,
- injection onto lunar trajectory $\Delta v = 3.15$ km/s,
- injection onto lunar orbit $\Delta v = -0.67$ km/s,
- landing on the Moon $\Delta v = -1.77$ km/s,
- launch from the Moon into lunar orbit $\Delta v = 1.77$ km/s,
- injection onto return trajectory $\Delta v = 0.67$ km/s,
- injection onto Earth orbit $\Delta v = -3.15$ km/s,
- deorbiting and braking $\Delta v = -7.91$ km/s,
- Total $\Delta v_t = 27.0$ km/s.

The speed increments needed to brake the vehicle are shown with a minus (−) sign. The addition is performed regardless of sign, as all manoeuvres use propellant. Because the final braking can be performed using aerodynamic drag (aerobraking), the total Δv for which fuel is required is appreciably smaller than the calculated value of 27.0 km/s.

If thrusters with an enhanced jet velocity are used, as when changing from chemical to electric propulsion, the thrust decreases and the propulsion time increases. The impulsive phases become longer and the coasting parts of the trajectory become shorter, up to a fully propelled trajectory with no coasting parts. The propelled arcs of the trajectory are similar to parts of a spiral and are more difficult to design (Figure B.4). The orientation of the thrust and its time profile become important in determining the total amount of propellant (and hence energy) used, and must be carefully and accurately optimised.

Continuous-thrust (or 'motorised') trajectories, i.e. trajectories for which the thrust is applied for a substantial part of the journey, are energetically less efficient than impulsive trajectories, i.e. trajectories when the thrust is applied in very short bursts, followed by coasting phases. The theoretical minimum expense of energy is incurred when

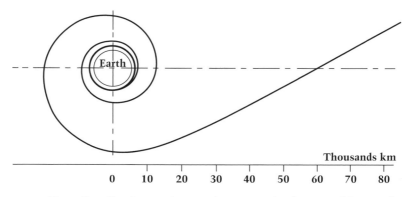

Figure B.4. Continuous thrust trajectory starting from an orbit around our planet; in this example it takes 92 days to reach escape conditions.

the Δv is imparted in a single short burst as close as possible to the starting planet.

Some numerical values of important parameters for astrodynamic calculations, for the major bodies of the solar system, are given in Table B.1.

LAGRANGE POINTS

Consider the system composed of one celestial body and a second body in orbit about it – to be more precise we should consider two celestial bodies orbiting about their common centre of gravity. The Earth and the Moon form such a system, as do the Sun and the Earth. In each such system there are five points, the so-called *libration points*, or *Lagrange points*, where a third body (a spacecraft, or a space station) is in equilibrium under the gravitational attractions of the first two. The Lagrange points for the Earth–Moon system, shown in Figure B.5, rotate around the Earth (actually around the centre of mass of the system) as the Moon moves around the Earth once a month.

Lagrange points are a solution of the so-called restricted three-body problem, in which it is assumed that the mass of the third body is much less than that of the first two spherical bodies.

The Lagrange points L1, L2 and L3, which lie on the line connecting the two main bodies, one between them and two outside them,

Table B.1. *Data on the main bodies of the solar system.[a] The diameter D is given in kilometres, the mass is referred to the mass of the Earth M/M_E ($M_E = 5.96 \times 10^{24}$ kg), and the density ρ is in kg/m^3, the mean radius of the orbit around the Sun (around the Earth for the Moon) R_0 is given in millions of kilometres, the orbital period T is in years, the gravitational acceleration on the surface g_0 is in m/s^2, and the escape velocity v_e is in km/s*

	D (km)	M/M_E	ρ (kg/m³)	R_0 (Gm)	T (years)	g_0 (m/s²)	v_e (km/s)
Sun	1,392,000	332,000	1,408			274	618
Mercury	4,876	0.05552	5,430	57.8	0.241	3.7	4.25
Venus	12,104	0.815	5,243	108.2	0.616	8.87	10.36
Earth	12,756	1	5,515	149.6	1	9.81	11.18
Mars	6,780	0.107	3,934	227.9	1.882	3.73	5.03
Jupiter	142,984	317.9	1,326	778.3	11.86	25.38	60.2
Saturn	120,536	95.2	687	1,432	29.37	10.44	35.48
Uranus	49,946	14.5	1,318	2,877	84.10	8.85	21.27
Neptune	49,528	17.1	1,638	4,498	164.9	11.14	23.49
Pluto	2,304	0.002	2,050	5,900	248.6	0.65	1.21
Moon	3,474	0.012	3,344	0.384	0.0748	1.624	2.38

[a]K. Lodders and B. Fegley, *The Planetary Scientist's Companion*, Oxford University Press, Oxford, 1998.

are points of unstable equilibrium. Thus, a spacecraft placed at any of these three points will drift slowly away from it under the effect of the slightest perturbation. To keep a spacecraft at such a position, a station-keeping system is essential. The solar observatory *Soho* placed at the L2 point of the Sun–Earth system proved that it is possible to stabilise a spacecraft there (*Soho* actually moves in a 'halo' orbit around the L2 point). A future large astronomical observatory could be located at one of these points (probably L1) in the Sun–Earth system.

Points L4 and L5 are at the vertices of two equilateral triangles, with the two celestial bodies being at the other vertices. These points

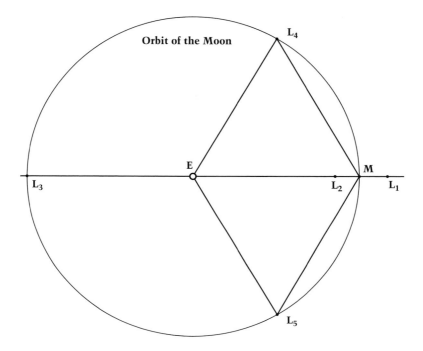

Figure B.5. The five Lagrange points for the Earth–Moon system, drawn to scale; the centre of mass of the system lies within the Earth.

are on the same orbit as the second body, leading it or trailing it by an angle of 60°. These points are stable equilibrium points, and a space-craft – or a natural object – remains there. A number of asteroids, named after Trojan and Greek heroes, are in such orbits in the Sun–Jupiter system, the Trojan asteroids are at L4 (leading Jupiter) and the Greek ones at L5 (trailing Jupiter). A space station could be constructed at either of these points in the Earth–Moon system.

NON-LINEAR ASTRODYNAMICS

In astrodynamics the gravitational force is a non-linear function of the distance from the major massive body; all astrodynamics is there-fore non-linear. The term *non-linear astrodynamics*, that branch of astrodynamics which studies the trajectories of celestial bodies and space vehicles taking account of their simultaneous interactions with several celestial bodies, is therefore something of a misnomer. The mathematical complexity of the subject of non-linear astrodynamics is such that the only possible approach is a numerical one. It is only possible at all due to the enormous computing power of modern digital computers.

The first non-linear effect to be exploited is the so called 'gravitational assist'. Here, a space probe increases its speed using the gravitational field of a celestial body, and at the expense of that celestial body's energy. The effect is much larger if the celestial body is large and fast in its motion around the Sun, and if the space probe gets fairly close to its surface. Jupiter, with its huge mass, and Venus, with its large orbital velocity, were both used in the *Galileo* mission. The Moon, although small, can be useful as its lack of atmosphere allows the space probe almost to skim its surface.

When performing a 'flyby' of a celestial body, a thrust may be applied when the spacecraft is closest to the planet, i.e. a 'perigee burn' (in the case of the Earth; 'perihelion burn' in the case of the Sun, etc.) is carried out. In this case the flyby is said to be 'motorised'. The increment of velocity is very effective in this case, as it is applied deep in the 'gravitational well' of the planet; the hyperbolic excess speed can be far larger than the Δv impressed by the rocket engine.

The most interesting application of non-linear astrodynamics uses 'fuzzy boundaries', five-dimensional surfaces in state space (which, since it is comprised of real space and velocity space, has six dimensions) where the gravitational attractions are perfectly balanced and the dynamics become chaotic. The trajectories of spacecraft in such a situation were first studied by Edward Belbruno, a mathematician then at the NASA Jet Propulsion Laboratory in California. He discovered that large savings in the amount of propellant needed for a mission could be made, even if at the expense of a longer flight time. Recently, it has been found that some comets are on orbits determined by 'fuzzy boundaries' existing near the orbit of Jupiter.

Nowadays, it is possible to steer a spacecraft through the solar system, exploiting its complicated gravitational field precisely and following extremely complex trajectories. Robotic space missions can then be performed using rockets far smaller than those required for Keplerian trajectories. The more limited the performance of the propulsion systems, the more complicated is the design of its trajectory; if very powerful thrusters were available the space probes could travel through the solar system almost in straight lines. The situation is similar to that occurring at sea – the skipper of a sailing boat needs to know exactly how to exploit the winds and currents, while the owner of an off-shore racing boat rushes across the sea using only the power of the engine.

RELATIVISTIC ASTRODYNAMICS

If the speed of a spacecraft could approach the speed of light, everything becomes more complicated. The theory of classical mechanics is no longer sufficient to design the mission – relativity plays an important role.

Relativistic effects due to gravitational fields are small; nevertheless, they are detectable in the orbits of celestial objects. For example, they are essential to compute the orbit of Mercury accurately, owing to its closeness to the Sun. They must also be accounted for in GPS systems, because the positions of the satellites must be known with great precision.

Relativistic effects depend on the ratio of the vehicle speed v to the speed of light c. The ratio v/c enters via the factor

$$\gamma = \frac{1}{\sqrt{1 - \dfrac{v^2}{c^2}}} \tag{B.9}$$

The factor γ exceeds unity appreciably only at speeds of more than half the speed of light (Table B.2).

Table B.2. *Values of parameters γ and $1/\gamma$ for some values of the non-dimensional speed v/c*

V/c	0.01	0.1	0.5	0.8	0.9	0.99	0.999
γ	1.0001	1.0050	1.1547	1.6667	2.2942	7.0888	22.3663
$1/\gamma$	0.9999	0.9950	0.8660	0.6000	0.4359	0.1411	0.0447

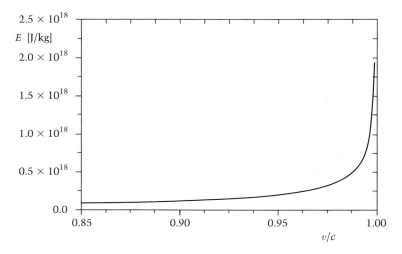

Figure B.6. Energy (in joules) needed to accelerate a mass of 1 kg as a function of the speed, as this approaches the speed of light c.

The relativistic contraction of time reduces onboard times by a factor of $1/\gamma$. The onboard time needed to travel a distance d at speed v which is approaching c is thus

$$t = \frac{d}{\gamma v} \tag{B.10}$$

This is the effective flight time for observers who are on the mission. The flight time as seen by an observer on the Earth is always $t_0 = d/v$.

The energy needed to accelerate a space vehicle, whose mass when at rest is m, to a very high speed v is, following the theory of relativity, given by

$$E_r = mc^2(\gamma - 1). \tag{B.11}$$

For $m = 1$ kg, this relation is plotted in Figure B.6. It is evident that the relativistic effects become crippling, in terms of energy, if v/c exceeds 0.99.

If the speed is much less than the speed of light, i.e. for $\gamma \approx 1$, equation (B.11) reduces to the well-known expression of classical mechanics for the kinetic energy of a body

$$E = \frac{1}{2}mv^2 \tag{B.12}$$

Appendix C The basics of space propulsion

ROCKET PROPULSION

A rocket produces a forward thrust by ejecting gas at high speed through a nozzle in the opposite direction. Both the material which produces the gas and the power source needed to accelerate it are carried aboard the rocket. In the case of a chemical rocket, the propellant performs both functions, using its chemical energy to accelerate the combustion product backwards.

The thrust T which a rocket exerts is given by Newton's second law of motion

$$T = \dot{m}v_e \tag{C.1}$$

where $\dot{m}$ is the rate at which the mass decreases due to propellant loss; v_e is the ejection velocity of the gas. The total impulse I which the engine supplies, when it operates with a constant rate of output of propellant, is equal to the thrust multiplied by the working time. This then equals the total mass of propellant m_p burnt in the engine multiplied by the ejection velocity ($I = m_p v_e$). The specific impulse I_s is often used to express the performance of a given propellant. It is defined as the impulse supplied per unit weight of the propellant

$$I_s = \frac{I}{m_p g_0} = \frac{v_e}{g_0} \tag{C.2}$$

where g_0 is the acceleration of gravity on the surface of the Earth, as conventionally the specific impulse is referred to unit weight on the surface of the Earth. The specific impulse is measured in seconds. It can also be interpreted as the time 1 kg of propellant lasts when burnt to supply a thrust of 1 kg force (about 9.8 N).

The specific impulse[1] is sometimes defined as the impulse supplied by unit mass (instead of unit weight) of the propellant; this coincides with the ejection velocity and is measured in m/s.

The propellant supplies both the energy needed to produce the exhaust gas and the ejected mass. Each fuel–oxidiser combination has its own ejection velocity, which increases with decreasing molecular mass of the products of the combustion and with increasing chemical energy liberated by the combustion

$$v_e \propto \sqrt{\frac{T}{\mathcal{M}}} \qquad \text{(C.3)}$$

where T is the temperature of the hot gases and $\mathcal{M}$ is their molecular mass. The constant of proportionality implicit in equation (C.3) depends on many factors, including the ratio between the pressure in the combustion chamber and the external pressure. Equation (C.3) holds for all thermal rockets, i.e. for all those rockets in which the jet is accelerated by heating the gas.

The hydrogen–oxygen combination used by the Space Shuttle engines has a combustion temperature of 2,400 °C and an ejection velocity of about 4600 m/s, the highest among the commonly used liquid propellants. The specific impulse of the engines is therefore almost 470 s. Such a value is, however, a theoretical limit, reached only when the expansion in the nozzle is complete. That would need an infinitely long nozzle and expansion into a vacuum. The actual ejection velocity is lower, and decreases with increasing ambient pressure.

Unfortunately liquid hydrogen cannot be used in all applications: apart from its low density, which necessitates large tanks, it must be stored at cryogenic temperature (below 20 K, i.e. minus 253 °C).

The velocity v_{ec} which can be reached by a rocket starting from standstill is

$$v_{ec} = v_e \ln \left(\frac{m_0}{m_{ec}} \right) \qquad \text{(C.4)}$$

---

[1] Further details are given in R.X. Meyer, *Elements of Space Technology*, Academic Press, London, 1999.

where m_0 is the total mass at launch and m_{ec} is the mass at the end of the combustion phase, i.e. m_0 minus the total mass of propellant; ln is the natural logarithm. The presence of the natural logarithm shows that the velocity of the rocket grows only slowly while significantly increasing the quantity of propellant used. To reach a speed equal to the propellant ejection speed, a rocket must burn an amount of fuel equal to 63% of the initial mass. To reach a speed twice that of the exhaust speed, 87% of the launch mass has to be fuel; this leaves a meagre 13% of the initial mass for the useful payload, the structure, and all ancillary devices.

Things are actually far worse than equation (C.4) indicates. When launching from the Earth's surface atmospheric drag slows down the rocket in the first part of the flight, and part of the thrust is used just to overcome the weight of the vehicle and does not accelerate it. For a launch in the vertical direction from a planet having an atmosphere, the final rocket velocity is

$$v_{ec} = v_e \ln \left(\frac{m_0}{m_{ec}} \right) - gt - v_r \tag{C.5}$$

where t is the time for which the thrust is applied and v_r is a term which accounts for the losses due to aerodynamic drag. It is clear that it is best to burn the propellant in the shortest possible time, but there are limits due to the maximum acceleration which the structure of the vehicle and the crew (if present) can withstand.

The power needed to produce the jet, and thus the thrust, is given by

$$P = \frac{1}{2} \dot{m} v_e^2 = \frac{1}{2} T v_e \tag{C.6}$$

For a certain value of the thrust, the power grows with increasing exhaust velocity.

As was seen in Chapter 3, a chemical rocket is a simple machine only in theory. This is particularly so in the case of liquid propellant rockets, where there are several ancillary devices to perform such tasks as pressurising the tanks, pumping and injecting the fuel and oxidiser into the combustion chamber, cooling the combustion chamber

Table C.1. *Comparison of the performance of various types of propulsion devices; for devices still at the research stage the values given are based on theoretical estimates*

Engine type		v_e(km/s)	I_s(s)	State
Chemical rockets	Solid propellant	1.8–2.5	180–250	in use
	Liquid propellant	2.4–5	240–500	in use
	Free radicals	21.3	2,130	basic research
	Metastable molecules	31.5	3,150	basic research
Nuclear rockets	Fission (solid core)	7–11	700–1,100	experimental testing
	Fission (liquid core)	13–16	1,300–1,600	basic research
	Fission (gas core)	17–70	1,700–7,000	basic research
	Fusion	25–2,000	2,500–200,000	basic research
	Pulsed	20–10,000	2,000–1,000,000	basic research
Solar rockets	Thermal	8–10	800–1,000	experimental testing
Electric rockets	Arcjet	20	2,000	experimental testing
	Plasma	30–50	3,000–5,000	experimental testing
	Ion	30–100	3,000–10,000	experimental testing

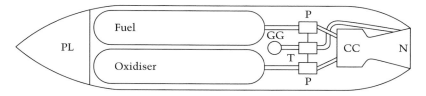

Figure C.1. Sketch of a single-stage liquid propellant rocket. CC indicates the combustion chamber, GG the gas generator, N the nozzle, P the pumps, PL the payload, and T the turbine.

and controlling everything. A sketch of a single-stage, liquid propellant rocket is shown in Figure C.1. Rockets with solid fuels are in principle simpler, but they suffer from very high mechanical and thermal stresses.

Instead of burning a fuel and an oxidiser, it is in theory possible to exploit the energy of metastable molecules or free radicals. The performance of the rocket would be higher, but so also would be the difficulties of producing and storing such chemicals. At present these types of rocket engine are still in the basic research phase.

The performance of various rocket types in terms of ejection velocity v_e and specific impulse I_s, is reported in Table C.1.

NUCLEAR ROCKETS

To obtain higher exhaust velocities, the functions of power generation and thrust production must be separated. In nuclear fission thermal rockets, a nuclear reactor produces energy which is used to heat a fluid; this is then expanded in a nozzle to produce the thrust. The simplest (yet difficult) method is to have a solid core reactor, as used in power stations (see Figure C.2).

The performance of all thermal rockets increases when the temperature of the jet is increased (equation (C.3)). In the case of solid core rockets, the core must not melt. The gas cannot reach a really high temperature, and the temperature is lower than for a chemical rocket. Solid core nuclear engines can reach high values of the specific impulse only because they can use a low molecular mass gas, typically

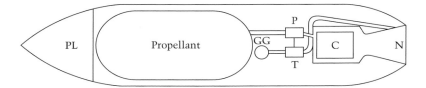

Figure C.2. Sketch of a nuclear rocket. C is the reactor core, GG the gas generator, N the nozzle, P the pump, PL the payload, and T the turbine.

molecular hydrogen ($\mathcal{M} = 2$) instead of water vapour ($\mathcal{M} = 18$) as in hydrogen–oxygen rockets.

To increase the temperature, a nuclear reaction in a liquid, or gas, core has been proposed. In particular, gas core reactors, in which the fissionable material is a gas, e.g. uranium hexafluoride, can overcome the intrinsic limitations of solid core reactors and yield substantially higher exhaust velocities. The difficulty is to prevent the core material from escaping with the jet; however, several designs for liquid and gas core rockets give promising results.

A nuclear rocket in which the gas (hydrogen) is directly heated by fission fragments (americium 242 is used instead of uranium 235 owing to its higher capture cross-section for neutrons) has been put forward by the Nobel laureate Carlo Rubbia. The advantages are a very high exhaust velocity, up to five times that of a solid core engine, and only small quantities of fissionable material – it is a very 'clean' nuclear rocket. The Italian Space Agency has started a project, called *Project 242*, to investigate this idea, with the aim of eventually building a spaceship to be used on the route to Mars.

Nuclear propulsion in general, and the '242 proposal' in particular, has the advantage of combining a high specific impulse with high thrusts. This is possible only because of the very high power supplied by the nuclear reactor. Nuclear engines are the best candidates for large interplanetary spacecraft, particularly for manned missions.

For safety reasons nuclear-powered rockets cannot be used for launching from the Earth's surface, at least until it has been shown conclusively that there is negligible leakage of radioactive material from the exhaust. There is no doubt that using nuclear technology would be a simpler way of building a single-stage to orbit (SSTO) vehicle than via conventional chemical rockets.

ELECTRIC PROPULSION

An electric current can heat a fluid which, after expanding in a nozzle, produces the jet. Electrothermal propulsion can be implemented either by using electric resistances (resistojets) or, better still, by using electric arcs (arcjets) (see Figure C.3a). An arcjet can achieve ejection velocities of about 20,000 m/s, almost ten times those with chemical rockets. Then the fuel needed for a given mission is much less. Obviously, the mass of the propulsion system is greater since, apart from the ancillary equipment and the chamber in which the fluid is heated, an on-board electric power generator is needed.

To obtain an even higher ejection velocity, thermal expansion of the fluid must be abandoned; the fluid must be directly accelerated by an electric field. The simplest scheme is to ionise the fluid and then to accelerate it using an electric field (ion engines, see Figure C.3c). Ejection velocities up to 100,000 m/s may perhaps be obtained in this way.[2] In the device shown in the figure (which is of the same type as that which performed flawlessly on the *Deep Space 1* probe) there is a grid acting as cathode that accelerates the ions; it must withstand intense bombardment by ions. In other devices, such as Hall effect thrusters, the accelerating electric field is produced by a stream of electrons orbiting around the axis of the device, in the magnetic field of a ring magnet. Hall effect engines, which were space tested by the Russians in the 1970s, have a lower efficiency than conventional ion engines, but have no grid and they are therefore likely to last longer. They may be scaled up to larger sizes, for manned missions. Very small ion thrusters, like the field emission electric propulsion (FEEP) device developed by ESA, can produce forces of a few millinewtons to control the attitude of satellites.

Generally speaking, ion engines are very low-thrust, high-specific-impulse devices.

---

[2] Science fiction has often used the concept of ion engines. The 'TIE fighters' of the *Star Wars* saga are 'Twin Ion Engines' spacecraft.

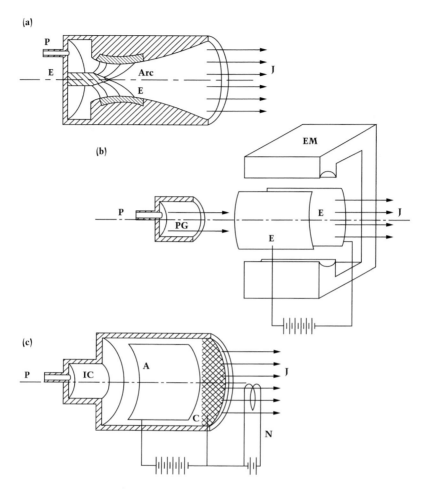

Figure C.3. Sketch of electric thrusters: (a) arcjet, (b) plasma engine, and (c) ion engine. A indicates the anode (focusing electrode), C the cathode (accelerating electrode), E an electrode, EM an electromagnet, IC the ionisation chamber, J the jet, N a neutralising electrode, P the propellant inlet, and PG a plasma generator.

In a magnetohydrodynamic (plasma) engine, the fluid is an electrically charged gas termed a plasma passing through a magnetic field while an electric current flows through it (see Figure C.3b). A force, of the same type as that acting on the windings of an electric motor, accelerates the fluid up to speeds of the order of 30,000–50,000 m/s.

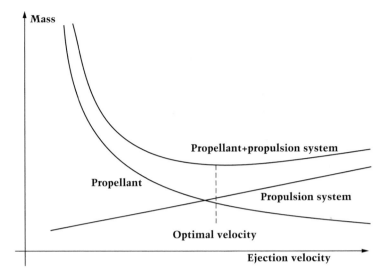

Figure C.4. Propellant and engine mass for a given space mission as functions of the ejection velocity. The optimum value of the velocity of the jet is shown.

The diagram in this figure explains the principle, but the actual layout may be quite different. Plasma engines can be scaled up better than ion engines, and might in the future be used as the main engine for a large spaceship.[3] A particular type of plasma engine is the Variable Specific Impulse Magnetoplasma Rocket (VASIMIR): the appropriate specific impulse between 1,000 and 30,000 s can be selected for the various phases of a spaceflight. Low specific impulse and high thrust (equivalent to selecting first gear in a car) is used to leave an orbit around a planet, while the engine can be put into 'top gear' to manoeuvre in deep space, where the gravitational field is very weak. NASA plans to flight test a small (10 kW) VASIMIR engine in 2004.

For small probes like *Deep Space 1*, fairly large solar panels can supply enough electric power. To constitute the main propulsion system for a large spacecraft, electric thrusters will need huge amounts

[3] Perhaps one of the best descriptions of a plasma engine is in the science fiction movie and novel *2001, A Space Odyssey*, by Arthur C. Clarke.

of electric power. Only a nuclear generator is adequate, and all the difficulties described in Chapter 3 will have to be overcome.

With increasing jet velocity the electric power needed to produce a given thrust increases, causing an increase of the mass of the entire generator. A decrease of the mass of the propellant is accompanied by an increase of the mass of the engine. Thus there is an optimum value of the ejection velocity for which the total mass of the engine plus that of the propellant is a minimum (see Figure C.4).

FUTURE PROPULSION TECHNOLOGIES

The propulsion devices described in the previous sections either already exist or could be developed within a relatively short time (more or less a decade). The relationship between ejection velocity (in m/s) and thrust (in N) for various propulsion systems is indicated in Figure C.5. The values shown are order of magnitude values, and some

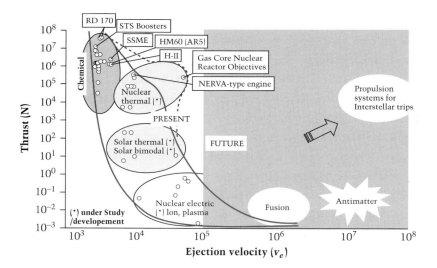

Figure C.5. Relationship between the specific impulse (here defined as the ejection velocity, and measured in m/s) and thrust of various types of space propulsion systems. (G. Genta, *Propulsion for Interstellar Space Exploration*, COSPAR Colloquium on the Outer Heliosphere: The Next Frontiers, Potsdam, Germany, July 2000.)

points are controversial, such as that labelled 'gas core nuclear reactor objectives'.

Some innovative concepts are described in Chapter 8. To master spaceflight and to go beyond the solar system, humankind has to develop new propulsion technologies which can yield both a high thrust and a high specific impulse. Powerful and compact sources of energy will also have to be developed. Propellantless propulsion, which would solve once and for ever the dilemma between high thrust and high specific impulse, requires not only technological advances but, first, a major breakthrough in theoretical physics. If, and when, that will happen cannot be foreseen.

Appendix D Common acronyms

Documents dealing with astronautics and space technology are often full of acronyms and abbreviations; some of them refer to the specific names of projects or missions. A list of some commonly used acronyms and abbreviations is reported here to help readers.

ABM	Anti-Ballistic Missile
ACS	Attitude Control System
AIAA	American Institute of Aeronautics and Astronautics
AL	Air Launched
APU	Auxiliary Power Unit
ASI	Agenzia Spaziale Italiana (Italian Space Agency)
ATM	Apollo Telescope Mount
ATP	Advanced Transportation Plan
AU	Astronomical Unit
BIS	British Interplanetary Society
BNSC	British National Space Centre
BM/C^3	Battle Management/Command, Control and Communications
BPP	Breakthrough Propulsion Physics
BSTS	Boost Surveillance and Tracking System
CETI	Communication with ExtraTerrestrial Intelligence
CFC	ChloroFluoroCarbon
CLEP	Conscious Life Expansion Principle
CNES	Centre National d'Etudes Spatiales (National Centre of Space Studies, France)
COPUOS	COmmittee for the Peaceful Uses of Outer Space (United Nations)
COSPAR	Committee for Space Research

C^4I^2	Command, Control, Communications, Computers, Intelligence, Interoperability
DARPA	Defense Advanced Research Projects Agency (USA)
DRL	Down Range Landing
ECLSS	Environmental Control and Life Support System
EDO	Extended Duration Orbiter
EO	Earth Observation
ESA	European Space Agency
ESOC	European Space Operations Centre
ET	External Tank
ET	ExtraTerrestrial
ETO	Earth To Orbit
ETI	ExtraTerrestrial Intelligence
EUMETSAT	EUropean METeorological SATellite organisation
EVA	ExtraVehicular Activity
FEEP	Field Emission Electric Propulsion
FESTIP	Future European Space Transportation Investigation Programme
FGB	Functional Cargo Block
FOBS	Fractional Orbit Bombardment System
FTL	Faster Than Light
GEO	Geostationary Earth Orbit
GLONASS	GLObal NAvigation Satellite System
GPALS	Global Protection Against Limited Strike
GPS	Global Positioning System
GTO	Geostationary Transfer Orbit
HST	Hubble Space Telescope
IAA	International Academy of Astronautics
IAF	International Astronautical Federation
ICBM	InterContinental Ballistic Missile
INTELSAT	INternational TELecommunication SATellite organisation
ISAS	Institute for Space and Astronautical Sciences (Japan)
ISS	*International Space Station*

KEW	Kinetic Energy Weapon
LEM	Lunar Excursion Module
LEO	Low Earth Orbit
LPS	Lunar Power System
MEO	Medium Earth Orbit
MMU	Manned Manoeuvring Unit
MOC	Mars Orbiting Camera
MOLA	Mars Orbiting Laser Altimeter
NACA	National Advisory Committee for Aeronautics (USA)
NASA	National Aeronautics and Space Administration (USA)
NASDA	NAtional Space Development Agency (Japan)
NASP	National AeroSpace Plane
NAVSTAR	NAVigation Satellite Timing And Ranging
NEA	Near Earth Asteroids
NEAR	Near Earth Asteroid Rendezvous
NERVA	Nuclear Engine for Rocket Vehicle Applications
NPB	Neutral Particle Beam
NTR	Nuclear Thermal Rocket
OA	Once Around
OME	Orbital Manoeuvring Engines
OMS	Orbital Manoeuvring System
OTV	Orbital Transfer Vehicle
PAH	Polycyclic Aromatic Hydrocarbons
PAM	Perigee Assist Module
PFC	PerFluoroCarbon
PHA	Potentially Hazardous Asteroids
RHG	Radioisotope Heat Generator
RLV	Reusable Launch Vehicle
rpm	revolutions per minute
RRL	Reusable Rocket Launcher
RTG	Radioisotope Thermoelectric Generator
RTL	Return To Launch site
RSTS	Reusable Space Transportation System
SAR	Synthetic Aperture Radar

S/C	SpaceCraft
SDI	Strategic Defence Initiative
SDIO	Strategic Defence Initiative Organisation
SETI	Search for ExtraTerrestrial Intelligence
SEI	Space Exploration Initiative
SI	Système Internationale (International System) of units
SLBM	Submarine Launched Ballistic Missile
SOSS	SubOrbital Single Stage
SNAP	Systems for Nuclear Auxiliary Power
SPOT	Système Pour l'Observation de la Terre
SPS	Space Power System
SRB	Solid Rocket Booster
SRM	Solid Rocket Motor
SSME	Space Shuttle Main Engine
SSTO	Single Stage To Orbit
STS	Space Transportation System (also used for the Space Shuttle)
TAV	TransAtmospheric Vehicle
TSS	Tethered System Satellite
TSTO	Two Stage to Orbit
VASIMIR	Variable Specific Impulse Magnetoplasma Rocket

Index